ECO

An essential sourcebook for environmentally friendly design and decoration

ELIZABETH WILHIDE

This book has been produced using part-recycled
fibre and/or paper sourced from trees grown in
properly managed sustainable forests.

Editorial Director: Jane O'Shea
Consultant Art Director: Helen Lewis
Project Editor: Hilary Mandleberg
Production: Tracy Hart

Illustrations: Vincent Westbrook
Picture Research: Nadine Bazar
Picture Assistant: Sarah Airey

First published in 2002 by
Quadrille Publishing Limited
Alhambra House
27–31 Charing Cross Road
London WC2H 0LS

This paperback edition first published in 2004

British Library Cataloguing-in-Publication Data
A catalogue record for this book is available from the British Library.

ISBN 1 84400 108 3

Printed in Hong Kong

contents

introduction

Eco design has come of age. Marrying time-honoured low-tech solutions with the latest in technological developments, contemporary eco houses are not merely environmentally friendly, but score highly in aesthetic terms as well.

Of all the dangers that threaten our world, damage to the environment is one menace we can all do something about. Superficially, this does not always appear to be the case. Despite the best efforts of green campaigners, international agreements to curtail global warming, limit deforestation and save natural habitats have so far fallen well short of the measures needed to address the situation the world finds itself in. When confronted with yet more evidence of the growing size of the hole in the ozone layer or a report that another species is on the brink of extinction, most people's response is a sense of powerlessness, partly to do with the sheer scale of the problem, but also because the issues themselves are complex and often contentious.

Each of us, however, can make a real difference; in fact, the best way of bringing about change may well be at the individual level. While experts argue and governments prevaricate, the choices we make as consumers represent direct action at its most immediate and effective. Collectively, those consumer choices, multiplied across thousands of households, can have a crucial impact in preventing and even reversing harm done to our planet. Think global, act local, is the environmentalist's watchword. Where better to start than in our own individual environments: at home?

Asked to imagine a significant agent of environmental damage and most people would picture a factory belching out toxic smoke, a tanker spilling gallons of oil or loggers ruthlessly destroying acres of rainforest. In fact, the greatest polluters on the planet are buildings. Buildings account for over half of total energy use in the developed world and produce over half of climate-change gases – air conditioning alone pumps more climate-change gases into the atmosphere than any other form of technology. What is equally serious is that buildings constructed today have a useful life of less than a century, which means more frequent replacement and greater waste. 'Building', of course, embraces a wide range of structures, from the out-of-town hypermarket to the office block. But the buildings we can all adapt and improve are the ones in which we live.

In recent years, a significant shift has taken place in the public's perception of environmental issues. Early advocates of the green cause tended at best to be viewed as eccentrics, a vociferous minority of doomsayers running helplessly counter to the forces of progression and modernity. To be 'green' was to be somehow cranky, alternative and generally New Age-ish; it was regarded as worthy, self-denying and often dull. But things have changed. Just as many people first tried organic produce after a food scare but kept buying it because it tasted good, we are at a similar turning point with eco design.

Increasingly, environmentally friendly design simply equates with best practice. It does not set itself against progress and technology, but often achieves green aims through highly sophisticated means. And while architects and designers have begun to explore alternative methods and materials, green campaigners have also begun to realize the importance of aesthetics. As the American writer Bruce Sterling has put it, the ultimate success of eco design will be recognized when it is 'more than politically correct, or even user-friendly'. It will happen when it is 'taken for granted' as part of the process of good design. When that day arrives, a building, product or material that fails on ecological grounds will be no more acceptable than a building, product or material that fails in some other structural, ergonomic or practical way. As the illustrations and case studies in this book demonstrate, good design and green design can and should be synonymous.

If it is not necessary to sacrifice one's aesthetic preferences to save the planet, neither is living 'green' all or nothing. The Hippocratic oath instructs medical practitioners to first 'do no harm'; in an eco context, this precept might usefully be rewritten as 'do as little harm as possible'. Building a house from scratch obviously provides the widest scope for incorporating ecological principles into the design process. But it is not always possible to reduce the environmental impact of buildings to zero. For those who can only adapt pre-existing surroundings, make a series of minor alterations or simply adjust their lifestyle, green choices and alternatives still offer an important opportunity to minimize damage done to the environment.

What is eco design?

Eco-friendly design is not just a question of choosing between good and bad alternatives, but involves complex equations that take into account both the broader picture and the longer view. While 'ecological', 'environmentally friendly', 'sustainable', 'green', 'natural' and 'organic' appear to be interchangeable concepts, they can imply quite different solutions. The fluidity of these terms has resulted in many interpretations by architects, designers and ecologists, and not a little controversy.

Broadly, ecological design is design that makes use of resources that come from the earth in such a way that they can be returned to the earth without causing harm, in a cycle that echoes the natural systems of living things. 'Sustainability', a related but not exactly equivalent concept, implies using resources, including land and energy, with maximum efficiency, at a rate that does not compromise the needs of future generations. While 'green' has become a blanket term for a range of environmentally friendly approaches, 'natural' and to some extent 'organic' are even less precise, particularly since such terminology has been increasingly appropriated by companies seeking to 'greenwash' their products. In design terms, 'natural' and 'organic' have also been used to describe buildings that echo the colours and forms of the natural world, which is not the same as designing to protect the natural environment.

ABOVE The Sustainable House in Leidsche Rijn, in the Netherlands, is a zero-energy detached villa that generates its own electricity from 50m² of solar panels on the roof. Other energy-saving features include thick insulation, double glazing and a heat-recovery unit. The house is part of a planned development of 30,000 new homes conceived along environmentally friendly lines.

Setting priorities

One of the central challenges in eco design is setting priorities: in other words, determining which efforts and strategies will do the most good. This entails both risk assessment on local and global levels and an appreciation of the specific opportunities provided by individual circumstances, factoring in issues to do with cost and effort.

In terms of risk, what the public identify as the most important environmental issues, often as a result of high-profile campaigns or news stories, are not necessarily those that feature high up on scientific assessments. An oil spill, for example, might attract attention disproportionate to its actual impact on the overall global environmental picture. In 1990, scientists working for the US Environmental Protection Agency produced the following ranking of environmental risk:

High-risk problems were identified as the destruction or alteration of habitats; the extinction of species, leading to loss of biodiversity; ozone depletion; and global climate change.

Medium-risk problems included the use of herbicides and pesticides; pollution of surface waters with toxins; acid rain; and air pollution.

Relatively low-risk problems were oil spills, groundwater pollution, acid runoff to surface waters and thermal pollution.

Risk also has a local dimension. Because eco design implies design that is in tune with specific climates and habitats – everything from hot arid desert landscapes to high-density inner-city sites – a fixed set of priorities cannot be produced that would be applicable in every case. In dry regions, for example, saving water would be high on the list of environmental aims; in other circumstances, an over-riding priority might be to safeguard the habitat of an endangered species. In general terms, however, sustainable building involves taking into account the following considerations:

Energy efficiency Buildings make their greatest impact on the environment through the energy they consume over their lifetime, so in most cases the first priority is to design and build for increased energy efficiency. Eco strategies that may be adopted include: improving insulation, specifying low-emissivity or other types of high-performance glazing, using renewable sources of energy, installing energy-efficient appliances, and designing and siting for passive solar gain/minimal heat loss (or the reverse in hot regions) and optimum volume-to-surface area ratio.

Material choice and use Buildings are vast consumers of materials. Eco strategies that may be adopted include: reducing material use and waste associated with construction; choosing materials that have low-environmental impact or are salvaged and recycled; and avoiding materials and decorative finishes with toxic constituents.

Site impact Eco design means sensitive land use: choosing brownfield over greenfield sites for new-build projects; minimizing disruption to existing trees and other aspects of the local habitat; and adopting sustainable approaches to gardening and landscaping.

Water use and efficiency The need to save water depends to a great extent on locale, but reducing water usage is likely to become a more critical issue in the future. Eco strategies that may be adopted include: using water-efficient plumbing systems and appliances; installing composting toilets; collecting rainwater; and recycling greywater.

Longevity and flexibility Buildings should not be regarded as disposable, nor should they be designed for obsolescence. By prolonging a building's effective usefulness its environmental impact is minimized over time. Eco strategies that may be adopted include: designing for durability, ease of maintenance and repair; enhancing flexibility so that a building can be adapted to suit future needs; and restoring or recycling existing buildings.

LEFT Bay of Fires Lodge, Tasmania, comprises a pair of pavilions that serve as an eco hiking lodge. It is fully sustainable with water collection on the roof, solar power and composting toilets. Only three trees were felled during its construction.

The benefits of eco design

OPPOSITE A demonstration
house on the outskirts of
London, Integer House
consumes only half the
energy of a comparable building
and cuts water use by a third.
A glazed conservatory bathes
internal spaces with natural
light – the ultimate feel-good
factor.

There's no getting away from the fact that eco design is fundamentally altruistic; it's investing in a future one won't necessarily live to see, on behalf of generations to come. But altruism, on the whole, is not a strong motivator. We live in a 'here and now' society and have grown accustomed to more or less instant gratification of our needs and desires.

Fortunately, eco design brings with it other benefits which can be more immediately persuasive. In fact, some self-builders of eco homes did not deliberately set out to design and construct on ecological grounds; they simply arrived at that point by opting for the most cost- and labour-efficient options.

In general, however, it is important to be aware that eco design does not necessarily save money in the first instance, nor cost less. While it seeks to reduce the use of materials, which can result in considerable savings for new-build projects, in some cases there may actually be a greater use of materials, for example thicker studs and rafters to accommodate thicker layers of insulation. While many eco homes cost next to nothing to heat and power and some actually earn income by exporting home-produced energy to national grids, at the same time, there are higher start-up costs associated with energy-efficient elements such as high-performance glass or solar technology, and the payback period for these can be as long as ten to twenty years. The economic picture, however, is not a static one; with increased demand, it is likely that many eco products or technologies may cost less in the future, as is already happening in the case of photovoltaics.

If the economic arguments in favour of eco design are not clear cut, one of the most immediately obvious benefits is improved personal health. Many modern materials and finishes commonly used in construction contain a huge number of chemicals and additives with a proven track record of causing illness. Chief among the culprits are volatile organic compounds (VOCs), a large class of chemicals that includes formaldehyde, organo-chlorines and phenols, which readily release vapours at room temperature or below. VOCs are present in carpets, underlays, paints, varnishes, vinyl flooring, insulating materials, seals and adhesives, household cleaners and air fresheners, among other products. Health problems associated with VOCs range from skin rashes, nausea, asthma and other breathing problems to chronic fatigue and dizziness; many VOCs are also suspected of being carcinogens. Designing and decorating with natural, untreated materials will help remove the threat caused by such toxins from the home.

Less tangibly but no less satisfying, eco design encourages an innate feeling of well-being and comfort. Natural light, fresh air and greater reliance on passive heating and cooling create environments in tune with biological rhythms – houses that feel like a third skin. Materials that connect with the land offer a quality of rootedness that gives a house a true sense of place, as well as providing unforeseen sensual delights, such as the velvety sound quality of a straw-bale house, for example.

Finally and perhaps most importantly, eco design is empowering. It often involves going against the grain of accepted practice, taking risks and assuming greater responsibility. When we are simply consumers at the end of a long and complex production chain, we are neither in charge of our future nor of our present. Eco design promotes a better style of life in the fullest sense of the term.

DESIGN

The essence of eco design is efficiency, both in terms of energy and use
of resources. The most elegant solutions arise when all the elements of
design and construction are properly integrated: in eco design, it's how
everything works together that really matters. In one sense, eco design
is a way of going back to first principles. In these style-conscious times,
it is easy to become obsessed with the latest trend and forget that the
purpose of a house is to provide shelter.

OPPOSITE Future System's earth-sheltered house on the fringes of a national park in Wales only minimally interrupts a natural site.

From the tropics to desert regions, temperate zones to the Arctic, climatic conditions around the world vary enormously. Human beings, however, can only survive within a fairly restricted range of temperatures and levels of humidity, and can comfortably tolerate an even narrower band. Houses are the means by which we protect ourselves from the elements.

In the recent past, houses have all too often been designed with scant regard for site or existing climate conditions, with the result that they have required the costly technological input of heating or air-conditioning systems to make them liveable. Go back further in history, however, before such technologies were invented, and other patterns of design are evident.

Vernacular styles of building developed out of a limited set of technological and material options. Houses were built (often self-built) of local materials because in default of transport links they had to be, and they were designed to accommodate the prevailing weather conditions because there was no other way of maintaining a reasonable level of indoor comfort. The varying forms of vernacular shelter, from the tent and yurt to the adobe hut or cob cottage, reflect a profound understanding of how to design structures that work within a given set of climatic conditions. In hot, dry areas, for example, houses are typically long and thin to dissipate the heat, with internal courtyards that set up patterns of cross-ventilation and increase the ratio of surface area to volume. In cold climates, many vernacular

houses have compact plans, with rooms tightly arranged around a central, heat-producing core. A traditional variation on this theme was for humans and animals to shelter together, with the animals on a lower level generating warmth for the humans on the upper. In areas where there is heavy rain or snowfall, vernacular structures have deep overhanging eaves that provide essential weather protection for exterior surfaces.

The passive design strategies enshrined in traditional building forms are an important source of reference for eco design. But they are not the sole point of departure. Contemporary eco design also looks forward, harnessing cutting-edge technology to maximize efficiency and reduce wastage. Among the most encouraging of such developments are current advances in photovoltaic systems which promise to usher in an era where houses, like the natural world, are solely powered by the clean, free energy of the sun.

While many of the examples illustrated in this section are new buildings specifically designed along ecological lines, a significant number of the issues addressed by these projects are equally relevant for people living in older houses, or those considering making spatial alterations to their existing homes. You may not, for example, be able to change the siting, orientation or basic construction of your home, but you can improve its insulation, install solar slates in place of conventional roofing or extend your home in such a way as to benefit from passive solar gain.

OPPOSITE Hope House, designed by Bill Dunster, combines energy efficiency with good looks. A three-storey glazed atrium maximizes solar gain; a massive concrete slab stores heat from the sun and releases it into the interior; fans direct warm air; insulation keeps heat in, and ducts, shutters and windows provide ventilation and internal temperature control.

Siting and orientation

Eco design is naturally site-sensitive. On both a micro and a macro level, the location of a building has a direct impact on its performance. The local ecology of the site, its gradient, orientation and exposure provide specific conditions, while the regional climate offers a more general context for design.

Designing with the sun and the climate

The ideal site for an eco house in the northern hemisphere would be a south-facing hillside, a location offering opportunities for both passive solar gain and thermal shelter. But cost and availability are obvious determining factors; many people, particularly those in urban areas, may find their options circumscribed by a shortage of viable alternatives.

Siting a house to benefit from passive solar loss or gain can dramatically reduce the amount of energy needed to heat or cool it. The first step is to determine the sun's strength, angle and path across the site, paying particular attention to any obstacles that might shade the site.

To take advantage of the sun's energy in cool climates, the house should face south (north in the southern hemisphere). The south side should be up to 60 per cent window, while the north-facing side should be more enclosed with minimal openings. Overhanging eaves will prevent the interior overheating in the summer, while large windows allow sunlight to penetrate during the winter. If it is impossible to site the house so that it faces due south, a deviation of up to 20 degrees will still allow the building to collect about 95 per cent of the available solar energy.

It also makes sense if internal arrangements follow suit, with living areas on the south-facing side of the house and service areas, such as kitchens and bathrooms, on the north. Because warm air rises, upside-down layouts, with living areas located on the first floor and bedrooms on the ground floor, will make the most of passive strategies.

The degree of exposure of a site also affects energy requirements. Since cool air collects in hollows, in cool climates the site should ideally be either level or slightly elevated, and major openings, such as doors and windows, should not face prevailing winds. Further heat loss can be prevented by setting the house back into a slope. Such earth-sheltered buildings take advantage of the 'thermal flywheel' effect: since the earth absorbs heat more slowly than the air but retains it for longer, an earth-sheltered building will be warmed by the ground when the weather turns cooler.

The bottom line of eco design is to adapt the house to its location. From this basic principle follows a range of different eco strategies:

• Siting and orientation based on micro- and macro-climatic conditions to maximize energy efficiency: framing the house with plants that can grow up to provide a living screen and help keep the building cool.

• Design and construction that involves the least possible disruption to the site: buildings that sit lightly on the land with minimal foundations; buildings designed around existing natural features, such as trees; buildings that leave natural habitats as untouched as possible.

• Blending the house in with its setting: using turf roofs to replace the area of ground lost in construction; sheltering the house in an earth bank or setting it into a hillside.

• Improving the quality of the site: choosing a brownfield site over virgin land; installing composting waste systems.

Where passive solar strategies are employed, the house must have adequate thermal mass to enable it to store the solar energy it has collected and release it back into the building's interior. Like the principle of the brick in the storage heater, internal walls and floors made of massive materials such as concrete, stone and brick, retain heat and release it gradually overnight when there is no solar gain.

In hot climates, sun is the enemy. Small openings on the sunny side prevent the interior from overheating, while low, sprawling house plans with open internal courtyards offer maximum surface area for cooling. Houses in hot climates can be sited to catch the summer winds. In tropical areas, houses that face prevailing winds and are elevated, either on stilts or on hillsides, maximize internal air movement.

OPPOSITE In hot climates, houses need ventilation and shade. In this Australian house generous openings, vegetation and a canvas awning provide all that is needed.

BELOW A low-energy self-build house in England has an unusual organic 'footprint' to accommodate the mature trees on the site. As a timber-frame construction, only minimal pad foundations were required.

Construction and form

In all parts of the developed world, construction is highly controlled. Building codes and regulations, regularly updated to ever more exacting standards, specify the materials and methods that are regarded as suitable for safe construction, particularly in terms of fire resistance and stability. In response to such legislation, the construction industry has typically reacted by enhancing the performance of structural elements: one example of this is the treating of lumber with chemicals to promote fire- and moisture-resistance. However, for anyone who is concerned about the environment, treating wood with potentially toxic chemicals to overcome such innate disadvantages is, of course, unacceptable.

Working within such stringent legislative frameworks is often a challenge for eco design. Construction methods or materials that are unproven in conventional terms may meet with strong opposition; some pioneering eco designers have found that before they could build the way they wanted to, they first had to campaign to change local codes.

Timber and timber-frame

Timber construction, which is very common for domestic building worldwide, makes good ecological sense. Wood is a renewable resource and has low embodied energy (see page 122); timber-frame structures are lightweight, easy to insulate to a high degree of effectiveness and can be readily converted, altered, added on to and remodelled. For the self-builder, timber-frame construction is also both cost- and labour-effective. And because timber construction is relatively lightweight, foundations can be minimal, hence some post-and-beam timber-frame structures rest quite simply on individual concrete pads, rather than on slabs, which means less disruption to the site.

Although timber-frame construction, disguised behind masonry or brick cladding, is very common in standard housing developments, all-timber construction is much more the exception in Britain and northern Europe today, because of the perceived risk of fire and the propensity of timber to rot when exposed to moisture. In some areas, timber construction is only permitted for structures of one or two storeys. Yet such restrictions may be a reflection more of poor standards of timber building than of the inherent qualities of timber itself. According to the Walter Segal Self Build Trust, a charity promoting self-building in Britain, over the last 30 years timber has been more exhaustively tested for its fire

performance than any other material. Such studies have revealed that in the sizes used in construction, timber is slow to ignite and, once ignited, burns very slowly. Treated with borax, a naturally occurring salt, timber meets British building regulations specifications for fire-resistance and spread of flame. Similarly, timber will not rot or suffer from insect attack if it is kept below 18–20 per cent relative humidity.

To achieve this, structures must be designed so that water runs off and air circulates freely. One ecological form of timber construction that achieves this, first developed at the Findhorn Foundation in Scotland, is the 'breathing wall' system. In this method, the cavity between outer and inner timber walls is insulated with cellulose fibre (made from

recycled newspaper) treated with borax to promote fire-resistance and ward off attack by vermin. Because there is no plastic vapour barrier, as there is in conventional buildings, the entire structure 'breathes', diffusing moisture very gradually so it is not trapped within.

Deep overhanging eaves help to protect wood-clad structures from water penetration but frequent re-staining and vigilant maintenance will be required. It is also important to use as durable a timber as possible. Some softwoods, such as larch, are nearly as durable as hardwoods, while Douglas fir is very resinous and therefore self-protective. Green oak is the best timber for construction and is also affordable. Whitewoods and redwoods should be avoided if possible.

Parts of the structure that might be more at risk from fire or vermin attack can be treated locally with borax paste squeezed into drill holes. Borax can also be applied externally provided it is subsequently stained over. An eco choice for external timber treatment is a non-toxic organic wood stain.

Because of their relatively low thermal mass, timber buildings do not store heat very well but can heat up quickly, while buildings with high thermal mass, which heat up slowly and lose heat slowly, require more energy to reach the desired level of warmth. But the low thermal mass of the timber building need not necessarily be a disadvantage. Such a building can be an ideal solution for modern working families: the heating can be left off during the day when

ABOVE Timber framing combined with timber cladding and brick provide an environmentally conscious building option.

ABOVE Adobe is an environmentally friendly, traditional form of house construction in some parts of the world.

the daytime and slowly release it into the interior at night, which makes them a good choice for houses that are designed with passive solar strategies in mind.

In both types of construction the raw material is earth. In adobe houses, earth with the possible addition of sand, straw and other binders is used to make bricks, either on site or commercially, and these can be formed into curved organic shapes. In the case of rammed-earth buildings, moist earth mixed with a small amount of cement or binder is tamped into forms where it sets hard; walls are typically rectilinear. Both adobe and rammed-earth walls must sit up on elevated foundations of concrete block or stone so that they are not eroded by standing water. Exterior stucco or rendering provides moisture-resistance.

Straw bale

The mere idea of building a house out of straw is guaranteed to raise the eyebrows of anyone familiar with the story of the three little pigs. Although getting such structures approved and off the drawing board has been a considerable challenge, an increasing number are now being constructed, particularly in the western United States.

In fact, straw-bale houses are as sound structurally as they are ecologically. Straw, the dried stalks left after grain crops are harvested, is a common agricultural waste product that more often than not ends up being burned and contributing to air-borne pollution; its re-use as a building material is therefore a good example of recycling. When used in construction, compacted straw bales work exactly like over-scaled building blocks, stacked flat in staggered rows and pinned together with thin steel rods to form load-bearing walls, or used as infill between members of a supporting structure, in which case they may be placed edge-on for walls of a slimmer profile.

Like adobe and rammed-earth walls, straw-bale walls need to rest on elevated concrete foundations, as wide as the bales themselves, to protect against moisture penetration. Once the walls are up, they are wrapped in wire to provide a key for the application of external stucco and interior plaster, which needs to be a breathable type to allow moisture to escape. The bales also need to be protected from the wet during the construction process. Bales may settle – up to several centimetres – so may need to be pre-compressed to prevent this from happening. Plumbing is also often run outside the structure to guard against the risk of leaks.

Straw-bale walls, up to 60cm thick, have excellent qualities of both sound and heat insulation. The particular

family members are out at work and switched on in the evening to more or less immediate effect.

Timber also has a role to play in hybrid types of construction. Good energy efficiency can be achieved when the basic structure is a lightweight timber frame, external walls are well-insulated timber and internal walls are made of high-mass materials such as concrete block or brick.

Adobe and rammed earth

A third of the world's population lives in houses constructed from one of the most ecological materials of all time: earth. Adobe and rammed-earth houses work particularly well in moderate and warm regions; in colder parts of the world the walls may need to be doubled up to provide an insulation space. The thick, noiseless earthen walls absorb heat during

soft, velvety acoustic quality of the straw-bale house is a characteristic much admired by enthusiasts of this form of construction. The density of the bales means there is insufficient oxygen available for combustion, so straw bales are naturally fire-resistant and, if properly sealed, they do not attract insects or other vermin.

Earth-sheltering and turf roofs

Earth-sheltering is an age-old vernacular tradition which offers considerable ecological benefits. Still common in many parts of the world where winters are extreme, such as Scandinavia and alpine areas, earth-sheltered houses exploit the ability of earth to modulate temperatures. Because earth gains and loses heat much more slowly than air – a time lag measured in months – indoor temperatures remain more or less stable year-round. This is known as the 'thermal flywheel' effect. The degree of earth-sheltering may vary, from structures almost entirely covered by the ground to those where only the roof and north-facing or exposed walls are sheltered and south-facing walls are open to the sun for solar gain. As well as being highly insulated, earth-sheltered houses are the ultimate integration of house and landscape.

A variation on a similar theme is the turf roof, which has become something of an eco symbol in recent years. Again,

like the earth-sheltered house, the turf or sod roof is a typical vernacular feature of nordic and alpine areas, where the thick layer of soil and grass adds extra insulation. Seeded with wildflowers and native grasses, grass or 'living' roofs directly replace land lost in construction, provide a habitat for local species and help to clean the air by absorbing carbon dioxide from it. Because the soil depth is relatively minimal, turf roofs do not need mowing.

Although a turf roof will provide some extra insulation and can help shield a house from rapid temperature swings in high summer, it does not form a critical part of an energy-efficiency strategy because the underlying structure of the roof needs to be well ventilated. How successful a green roof will be will also depend on local climate. In areas of heavy rainfall, 'living' roofs will blossom in spring and stay green all year round. In drier areas, even when they are planted with the hardiest grasses and are well mulched, such roofs die back in the summer and are at their best only during the spring months. Because the grass is under stress for most of the year, weeds can easily take over and, in severe cases, the turf may erode in places.

As is the case with earth-sheltered houses, the structure of a turf roof needs to be stronger than that of a normal roof and it must be properly waterproofed.

ABOVE Turf roofs are not only good insulators, they also replace areas of habitat lost in construction. Seeded with native grasses and wildflowers, they are very low maintenance.

Ventilation

Like natural light, fresh air is a feel-good factor of the first order. But fresh air is not merely clean air; it is air on the move. In temperate regions, there is the need to strike the right balance between air movement, which contributes to comfortable, healthy indoor conditions and which lowers levels of humidity, and potential heat loss, which wastes energy. The resultant strategy is often summarized as: 'Build tight, ventilate right.'

In conventional design, air movement is often promoted mechanically with fans and extractors, while in some parts of the world, notably the United States, cooling is increasingly achieved by air conditioning. Air, however, moves naturally according to pressure differences and, in addition, warm (and polluted) air rises and cool air falls – the 'stack effect'. These facts can be exploited to provide passive ventilation and cooling strategies for both indoor and outdoor air that can achieve optimum comfort largely without resort to mechanical or technological systems.

The traditional method of promoting air movement indoors is through cross-ventilation: opening doors or windows opposite one another to draw air through the interior. In hot countries the siting of these openings can be extremely crucial. Windows and doors aligned with the prevailing winds make use of the pressure difference between the side of the house that faces the winds and the other 'wind-shadowed' side. The result is that cooling air is drawn through the interior spaces.

The stack effect can also be exploited by venting warm air at upper levels. High-level windows, skylights or passive thermal chimneys will draw warm air upwards through the

ABOVE In this community centre in Yirrkala, Australia, designed by Glen Murcutt, slatted panels made of local timber provide natural ventilation at the same time as screening from strong sun.

levels of a house. Window design is also important for achieving a good flow of air. Louvred or shuttered openings, typical in hot regions, allow the airflow to be adjusted according to the time of day.

In temperate or cool regions, where energy conservation is important, openings are increasingly tightly sealed to prevent warm air escaping, but this can pose a problem when it comes to natural ventilation. However, as long as the walls are not allowed to cool down excessively, opening the windows a few times throughout the day for short periods of time to refresh the air will not compromise energy efficiency significantly. However, areas of the home that are naturally more humid will require extra ventilation to prevent the build-up of condensation. Today, basic fans in bathrooms and kitchens and frame or trickle vents on windows are required for all new houses. The fans can simply be used when needed or may be controlled by humidistats or timers.

Another strategy is to fit a heat exchanger or heat-recovery ventilator as part of a mechanical ventilation system. These recover heat from vented warm air and transfer it to fresh cool air coming in, but such systems are expensive both to install and run and require ducts to be installed throughout the house. An alternative is passive stack ventilation using thermal chimneys or flues to ventilate kitchens and bathrooms without the use of fans. It is not cheap to install but is free to run.

Cooling

One direct way of cooling a house in a hot climate is to reduce the amount of solar energy that falls directly on the building or that enters it via windows. Deep overhanging eaves keep the sun off exterior walls and windows, as do verandahs, brises-soleil, pergolas and covered porches. Openings on the sun side should be few and small in size, or screened and shuttered for light and heat control. Reflective or white external finishes also serve to reduce the amount of heat absorbed by the building. Vegetation next to the house – trees, creepers and climbers – is another way of providing shade: deciduous species will shade walls in temperate climates during the summer, but will allow maximum sunlight through in the winter. In general, planted areas around the exterior of the house will have a more cooling effect than hard, light-reflective surfaces.

In hot regions of the world, the inner courtyard is a feature of many vernacular building types. It cools by maximizing the surface area available for heat loss and by facilitating patterns of cross-ventilation. Another traditional

feature, common in the Middle East, is the wind scoop or wind catcher, a flue mounted on the roof that channels air down through the building in a masonry channel, cooling it along the way.

Air that has travelled over water is naturally cooler, something that accounts for the age-old practice in hot climates of damping down pavements, terraces or courtyards immediately adjacent to living spaces in order to provide a fresh, cooling breeze. Sunken garden or courtyard pools will achieve the same effect.

ABOVE Cool air travelling over water plus sliding doors help provide natural ventilation in this house in a suburb of Melbourne, Australia.

Energy efficiency

Energy efficiency, which embraces a range of strategies, from basic draughtproofing to using energy-efficient appliances and heating systems, is key to eco design. Of these strategies, one of the most critical is insulation. The colder the climate, the more important it is for a house to be well insulated. Heat is lost primarily through walls, windows, the roof and the basement (if any). It is estimated that merely by insulating walls and the loft space, heat loss could be reduced by half.

Materials vary widely in their ability to conduct heat, with metals, for example, being very efficient heat conductors and hence poor insulators, and light porous materials, like wool, very poor heat conductors and consequently good insulators. Air is also a very poor heat conductor, which is why materials that are honeycombed with air pockets are good insulators. Increasing the thickness of a material also increases its insulating properties.

How well a structural element, such as a roof or wall, performs as an insulator is expressed as its U-value (or heat transmission coefficient), a figure which is derived according to a formula that takes into account the thermal conductivity of each of the components that makes up that element. For example, in the case of a standard cavity wall, the calculation is based on the conductivity of the exterior brick, the airspace, the insulating material, the interior blockwork, and the plaster or other finish. The lower the U-value, the higher the degree of insulation provided.

The standard way of insulating is to line roof spaces and walls and fill cavities with a bulky insulation product. This may be made of a number of different materials, including cellulose, mineral wool, glass fibre and extruded polystyrene, some of which are more environmentally friendly than others. In most cases, ground and basement floors also need insulation, as do water tanks and pipes. Heat loss can also be reduced by using a highly reflective material, such as foil, to bounce the heat back across a void or cavity instead of the heat being absorbed by a wall – a conventional example is placing foil behind a radiator on an external wall so that heat is directed back into the room.

Insulating materials

Choice of insulating materials is a thorny issue for the eco designer. Unfortunately, the most efficient types of insulation from the point of view of energy conservation are synthetic materials, whose production is particularly environmentally damaging and whose presence in buildings has been strongly

linked to health risks. These include polystyrene panels and foam, and polyurethane panels and foam. Polyurethane panels are twice as efficient as other insulators. A more eco-friendly alternative, although still a synthetic, is insulating blanket made of polyester, which can be recycled and is made of recycled materials. Unlike polystyrene and polyurethane insulating products, polyester does not require chemical treatment to make it fire- or pest-resistant.

The most common insulating materials are made from mineral wool. Again, there are question marks about their safety, particularly in terms of human health. Mineral-wool insulation – typically stone wool or glass wool – is thought to pose a similar risk to asbestos, which has led to the development of new glass-wool products in which the fibres are stronger and less likely to break off and be inhaled.

Among eco-friendly alternatives to these efficient but problematic products the most widely available is cellulose, which has a similar U-value to mineral wool. Cellulose insulation is made from recycled newspapers (not glossy magazines), mixed with a number of additives, including borax, to make it resistant to fire and mould. It is available as a loose filler and in panels, or it can be sprayed in a slightly dampened state directly onto walls. Advantages of cellulose insulation include the fact that it allows moisture to be evenly diffused, so making it ideal for 'breathing wall' systems (see page 22). It is also a good filler for cavities in walls, floors and roof spaces. Sprayed cellulose is tight-fitting and reduces air leaks considerably. Because it is important to achieve the right density, specialist contractors should carry out the work.

Other recommended 'green' insulating materials include various types of wood fibreboard, sheep's wool and panels and rope made of flax. The flax used for insulation consists of waste fibres from linen production. Flax rope can be used as a substitute for polyurethane foam. Sheep's wool, an excellent insulator, used to be an expensive option but has recently come down in price. Unlike mineral wool, which permanently sags and thins if it gets damp, sheep's wool recovers its natural springiness.

Preventing cold bridges

Heat always takes the path of least resistance, travelling readily through materials like metal that are good conductors. A cold bridge occurs where there is such a pathway between the cold exterior wall of a building and the interior surface of that wall, with the result that the inner surface drops in temperature. Metal ducts and grilles, metal window frames, single-glazed windows, solid masonry walls

OPPOSITE Insulation is the key to energy efficiency in cold climates. This highly insulated converted barn in Lech, Austria, is constructed of local pinewood.

LEFT New eco insulating materials are emerging all the time. This natural fibre insulation is composed of 95 per cent post-industrial denim scraps which would otherwise go to waste, combined with a small percentage of synthetic filler and borates to promote pest- and fire-resistance. It is denser than standard fibreglass insulation and a better soundproofer, but costs a third more.

LEFT Windows should be tight-fitting and at least double-glazed to prevent excessive heat loss. Very high-performance windows are triple-glazed with low-E glass.

and solid metal or concrete lintels are some common forms of cold bridge. Cold bridges are also more likely to occur in the corners of buildings, where two surfaces are exposed to the outside air. Cold bridges should be avoided wherever possible not only because they drain heat from the interior but also because they will cause condensation, which can then lead to the growth of mould. Avoiding them entails designing details, such as frames and lintels, in separate exterior and interior sections, installing double or triple glazing, and increasing external insulation.

Draughtproofing and windows

Basic draughtproofing is an important element in energy conservation. Most people are aware of the need for windows and doors to be as tight-fitting and airtight as possible, but the average home offers many more potential sites for the leakage of heat, including gappy floorboards, disused fireplaces, air vents and ducts. All of these should be stoppered and sealed up wherever possible. To minimize draughts from windows and doors, curtains made of heavyweight fabric and which are also lined can be used. In cold climates, an intermediary entrance space, such as a porch, connecting the main entrance with the rest of the house, is a useful way of providing a buffer zone that will help minimize heat loss.

Ideally, windows should have the same thermal performance as the adjacent walls in order to prevent cold bridging and condensation (see above). Double-glazing, whereby a layer of insulating air is trapped between outer and inner panes of the double-glazed unit, is a standard means of improving thermal performance.

Recently, some new types of glass have been developed for even greater energy efficiency. Low-emissivity or low-E glass incorporates an insulating coating that manages to keep heat loss to a minimum and that effectively serves as a form of thermal insulation. Low-E glass is particularly valuable for extensive areas of top lighting, such as glazed roofs, which under normal circumstances would produce uncomfortable extremes of indoor temperature. Low-E double- or even triple-glazed units, where the gap is infilled with argon or krypton gas, are very high-performance windows. Some of these units have blinds fitted between the panes – an arrangement which also keeps the blinds clean. The blinds are operable from the inside so that heat gain may be controlled in warmer months.

Using thermal mass

Heavy, massive floors and walls made of materials such as stone, concrete and brick which heat up slowly and retain heat for long periods can be used almost as storage radiators, absorbing heat during the day and releasing it overnight. In earth-sheltered buildings (see page 25) even longer 'thermal flywheel' cycles, measured in months, rather than days or hours, come into play.

In hot climates, the same principle can be exploited to cool the building. At night, windows and vents are opened to cool the mass of the building so that the next day more heat can be absorbed.

RIGHT Solar panels discreetly tucked away in a corner of the garden help provide a house with its electricity requirements.

OPPOSITE New developments in solar energy include solar slates and tiles which are much more affordable than previous forms of solar technology.

Heat and power

A key aim of eco design is to rely as little as possible on supplementary energy, particularly energy from non-renewable sources such as coal and gas. When a building has been designed with passive solar strategies in mind and constructed with a high degree of thermal insulation it should, in theory, require little in the way of additional heating, except on very cold days.

Ideally, that heating, as well the domestic power supply, should come from a renewable source, such as wind, wave, tidal or solar energy. The easiest way of achieving this is simply to sign up to a green provider; in Britain, for example, RSPB Southern Electric offers guaranteed hydro-powered electricity for the same price as non-green electricity. Wind-generated power from wind turbines is another form of

renewable energy that has been explored in recent years; the current world leader is Denmark, which generates 15 per cent of its electricity that way. Despite the obvious advantages of wind-generated power, there are associated difficulties. Large wind farms can be noisy, and many people regard them as an eyesore. More worrying, from an ecological point of view, is that wind farms sited near migration routes have been responsible for killing significant numbers of birds.

Solar energy, on the other hand, requires no large-scale installation, and the technology has advanced significantly over the past few years. Solar energy systems are now capable of meeting a high percentage of domestic needs for water heating and running appliances.

Less ambitiously, installing a new central heating system

and electronic controls can also provide considerable energy savings, particularly when combined with insulation. In some parts of the world, grants are available for those seeking to make energy-efficient improvements.

Improving central heating

If you have no option but to rely on conventional fuel, such as oil or gas, for heating, you can at least ensure that your consumption is as low as possible. Firstly ensure that your central heating system is scaled according to your household's need and the level of insulation; very big systems are wasteful. In addition, hot water pipes should be insulated and it can be a good idea to fit radiators with individual thermostats to provide precise heating control in different areas of the home. However, individual controls can be inefficient in terms of temperature swings, which can be wasteful, and because each thermostat can 'call' on the boiler individually, the boiler spends more time 'on', responding to each thermostat on demand.

Another way of ensuring low energy consumption is to invest in a better central heating system. If your boiler is between ten and fifteen years old, it is worth considering replacing it. Technological improvements mean that boilers and control systems today are many times more efficient than they were even five years ago, now achieving efficiency ratings of more than 90 per cent. The most efficient are condensing boilers, which reclaim heat from exhaust gases, and their efficiency is enhanced by modern thermostats and electronic controls. 'Boiler managers', devices that effectively learn patterns of consumption so that heat is not produced unnecessarily, can also be a worthwhile investment.

There is now a recognized need for very efficient low-output boilers that can provide sufficient hot water for the average household and are designed for well-insulated houses. At present, the options available include instant water heaters with secondary coils for space heating, and boilers that run at the low temperatures suitable for underfloor heating, both of which are currently more costly than standard systems.

Solar energy

One form of energy which is clean and infinitely renewable is energy from the sun: the world's entire energy requirements at any given time could be met 10,000 times over from solar energy alone. Even Britain, where cloudy days are commonplace, receives 750 times more energy from the sun than its annual electricity consumption.

ABOVE A solar panel on the roof of a glazed conservatory proves that looks do not have to be sacrificed for practicality.

Solar technology has improved massively over the past few decades both in terms of efficiency and aesthetics. It has also plummeted in cost. The relevant technology is fifty times cheaper than it was thirty years ago and five times cheaper than five years ago. Replacing an ordinary roof with a solar roof is currently about four times as expensive as replacement with the original type of roof, but the price is lower for solar roofs that are installed when a house is built. Solar roof prices are expected to drop dramatically in the coming years. There are some lending institutions that already provide solar mortgages at standard rates. It is important to be aware, however, that the payback period is measured in decades – up to twenty years at present – which means you need to take the long view. However the real case in favour of installing a solar roof is the resulting reduction in the production of carbon dioxide: solar energy is clean.

There are two main ways in which energy is actively, rather than passively, harnessed from the sun. The first is a thermal solar energy system, which generates heat, and the second is a photovoltaic system, which generates electricity.

Thermal solar energy systems

Easy to maintain and adaptable, these systems generally use solar collectors to heat water or air which is then distributed via ducts and vents to where it is needed. Collectors can be fitted to existing roofs, or can be installed on adjacent structures or in the garden; the location and angle of the solar collectors is critical.

Such systems are ideal for supplying hot water for houses in warm climates, and they are an excellent way of providing warm water for swimming pools. In temperate or cooler regions, however, they are best used as a means of preheating water that can then be brought up to the right temperature with less energy than would otherwise be needed. But the disadvantage of a thermal solar energy system is that it produces most heat when demand is lowest and the only way of countering this would be a space-devouring and hugely expensive means of heat storage.

Photovoltaics

Photovoltaic collectors convert radiant solar energy into electricity that can be stored for later use. In some areas, there is the potential to hook up to the national power grid, so that surpluses of domestically produced solar energy build up as energy credits to be drawn upon during times of heavy demand or low supply; some solar-powered homes even earn a modest income. But the ecological savings are more persuasive. An average solar-powered house saves a tonne of carbon dioxide a year; during its lifetime, the savings are as high as 34 tonnes.

The basic component of a photovoltaic system is the photovoltaic cell. This is largely made of silicon, a plentiful material that makes up 28 per cent of the earth's crust. Photovoltaic systems are modular and can be adapted for a wide range of applications. The most familiar format is the solar panel, which is typically installed on a roof as a bolt-on element. The panels should be sited in order to have the maximum exposure to sunlight. They perform best in clear conditions but will even work when it is cloudy, which is an advantage for those living in less sunny climates. Panels have a working life of over twenty years; after between two and five years they will have produced as much energy as they took to manufacture.

LEFT Wood-burning stoves make an attractive feature in a room but are best as a supplement rather than as the main heat source for a home.

Other, newer, integral formats of photovoltaic systems include solar tiles and slates, which come in a variety of colours, from yellow and red through to blue, green and purple. There are also translucent cells which can be installed over glazed roofs or incorporated into windows.

Designing photovoltaic systems is a complex business. They produce direct-current electricity, which in most cases will need to be converted to alternating-current electricity. This is done by means of a device known as an inverter. Systems that are connected to the national power grid use the grid as a means of storage; other systems use batteries and some use both.

It is important not to view photovoltaics as an ecological band-aid. For optimum energy efficiency you may need to make changes to your lifestyle and energy use, for example by reducing the number of appliances in your home, by swapping old appliances for newer, more energy-efficient models, and managing activities that demand high levels of power to avoid unnecessary peaks.

Wood-burning stoves

Developments in wood-burning appliances and stoves mean that one of the oldest forms of heating is now surprisingly environmentally friendly. Wood is a renewable resource, unlike coal or gas, and new clean-burning systems minimize smoke emissions while maximizing delivery of heat. Traditional European-style tiled wood-burning stoves also have the advantage of storing heat for a very long time. Overall, however, wood-burning stoves are not an efficient way of heating hot water unless they are fuelled with 'found' wood and so are best reserved as a means of supplementary heating in extreme conditions.

Natural and artificial light

The quality of light is intimately bound up with our sense of emotional and physical well-being. Natural light stimulates the production of vitamin D and regulates hormone levels; our biological rhythms are attuned both to daily and seasonal changes of light levels.

Light airy indoor conditions support a whole range of everyday activities. On eco grounds, the more daylight the interior receives, the less you will need to rely on artificial lighting and hence the lower your energy consumption will be. At the same time, sunlight means heat gain. In the design and placement of windows a balance must be struck between solar gain and heat loss.

Windows

Optimum window placement and internal layouts for passive solar strategies in temperate, cold and hot climates have already been outlined (see page 26). Sometimes, however, there is the need to introduce light into areas of the home which receive little direct sunshine. Windows placed high up in a wall allow light to penetrate further than windows at normal height; similarly, raising the height of a window, rather than widening it, brings more light in. Top lighting, via windows in the roof or skylights, can spill light down through connecting areas such as halls and staircases which might otherwise have no direct lighting at all. Internal windows, or

openings in partition or solid walls, allow 'borrowed' light from rooms which are directly lit through to areas with no windows or with a poor level of natural illumination.

Low-energy lighting

Excluding power used for heating and hot water, artificial lighting accounts for about 10 per cent of energy consumption in the average household. Considerable savings can be made by replacing standard light sources with energy-efficient ones.

The most common source of artificial light is the familiar tungsten bulb, a design that has changed relatively little since Edison's day. We are used not only to its warm, flattering tones but also to its shape. However, these bulbs are particularly inefficient, converting only 5 per cent of the energy they use into light, with the remaining 95 per cent converted into heat – a fact readily appreciated by anyone who has ever scorched their fingers on a lit bulb. Energy-saving or compact fluorescent bulbs, on the other hand, produce six times as much light as an ordinary bulb, which means wattages and power use can be dramatically lower. They also last up to eight times longer (8,000 hours as opposed to 1,000 hours), a factor which must be offset against their relatively high cost. While recent improvements have reduced the flickering previously associated with this type of light source, the colour cast is still not particularly comfortable. In areas where concentrated work takes place, full spectrum or daylight bulbs may be preferable.

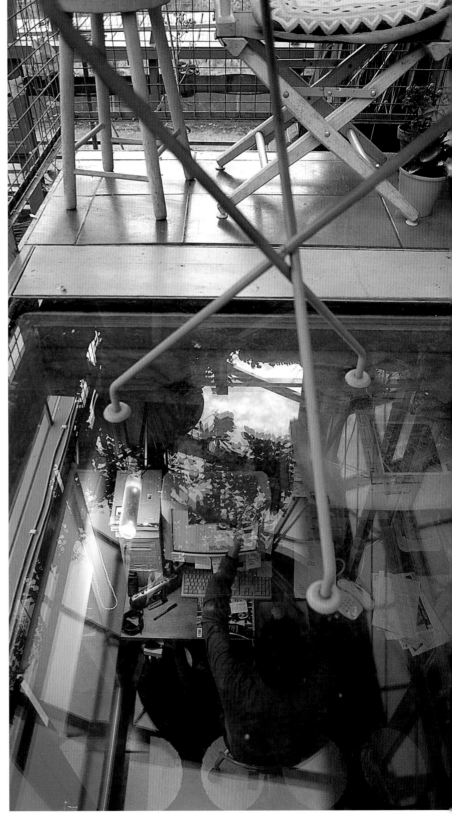

ABOVE Light pours right through an earth-sheltered home in the heart of the Welsh countryside.

RIGHT A glass floor allows light to flood through from an upper level to a lower one where the light would otherwise have to be supplemented by electricity.

Water and waste

Fresh water is one of the world's most precious resources and one whose supply is increasingly affected not only by pollution and global climate change but by rapidly escalating demand. Over the last fifty years, water consumption has trebled, but only a small percentage is used for drinking. In the home, the greater proportion is used for washing, bathing, dishwashing and flushing toilets.

In most cases, water consumption can be reduced significantly by adopting simple lifestyle changes, (see 'In Practice', page 166). Such strategies range from opting to shower rather than bathe, to installing water meters. Many owners of eco homes, however, go much further in the quest for full autonomy.

Rainwater collection

Collecting rainwater and storing it in large tanks or cisterns is an age-old means of ensuring a domestic water supply. Rainwater is not potable, but it can be used to irrigate gardens and wash cars, and even flush toilets and in washing machines provided it is treated to remove impurities which might otherwise damage plumbing. However, sophisticated systems incorporating filtering and purification tend to be prohibitively expensive.

Rainwater collection entails catching stormwater that runs off the roof, which means the roofing material is critical: many common roofing materials, such as asphalt and lead, will taint the water unacceptably. In built-up areas, there can also be significant pollutants in the rainwater itself. Aside from some means of filtration, there will also need to be a storage container of considerable size, ideally located underground, which in most cases needs to be connected to a pump to deliver the rainwater supply. There must be no connection between rainwater and drinking water pipes because of the risk of contamination. Rainwater taps should be clearly labelled.

Greywater systems

Greywater is water that is mildly soiled from washing, bathing or showering, as opposed to 'blackwater' which is water flushed from toilets as well as that from kitchen sinks, dishwashers and washing machines, which also counts as blackwater due to its high level of contamination from detergents, grease and organic matter. Greywater cannot simply be collected and re-used; it must first be filtered and treated or there will be a risk of disease and plumbing

blockages; after treatment it can be used for garden irrigation and to flush toilets.

Systems to recycle greywater have become increasingly prominent in recent years. In these, greywater is first coarsely filtered through crushed stone or gravel, then run through reed beds to be biologically purified by micro-organisms feeding among the roots of the plants. Such systems need to be carefully designed and sited to avoid pollution by groundwater and, of course, are very demanding in terms of land. Mechanical 'in-house' filtration systems also exist.

Composting toilets

One dramatic way of cutting water consumption is to install a composting toilet – a waterless system that breaks down human waste matter into organic compost. The toilet is connected by a shaft directly to a large sealed container in a basement or lower level. Air circulates through the container to break down the waste matter and an exhaust vent extracts smells and emits them above the roof. There may also be a shaft for adding organic waste from the kitchen and another opening for adding garden waste. Large containers only need to be emptied every couple of years. Although composting toilets have increasingly been adopted in some parts of Scandinavia, such systems are currently prohibited in many urban areas of the United States.

ABOVE A composting waterless toilet cuts water use and pollution.

RIGHT Waste from a composting toilet is collected in a large container, in this case located in a basement level. Eventually, the waste can be removed and dug into the garden soil, providing completely free, nutrient-rich compost.

Gardening and landscape design

For most of us, our garden is our immediate point of contact with the natural world. The difficulty with many domestic gardens, however, is that they have evolved to such an extent that they are now highly unnatural places.

Today many gardens, no matter where they are located, are dominated by the monoculture of the lawn, from which all other plants are ruthlessly excluded. The lawn, in turn, is fringed or interspersed with flower beds, borders and areas of shrubbery which are just as likely, if not more likely, to feature exotic plants from halfway across the world as they are local species. To maintain such gardens at the peak of perfection often entails constant irrigation and liberal doses of chemical insecticides, fertilizers and weedkillers. Many people today worry, rightly, about the use of such chemicals in the production of our food crops. Yet the concentration of chemicals in the average domestic garden can be many, many times greater. At the same time, gardening is sometimes no

longer viewed as a worthwhile and pleasurable physical chore; instead there is greater reliance on energy-hungry mowers and tools. A petrol mower running for an hour pollutes as much as a car driven 560 kilometres.

The eco approach to the landscape is firmly rooted in the locality itself. Rather than imposing an artificial garden ideal on a given site no matter what the prevailing conditions, the fundamental strategy is to work with the site to promote native species, both plant and animal, to conserve water and energy, and to employ organic methods of control. Those aiming for full sustainability will also grow their own fruit and vegetables for the table.

Working with nature

Eco gardening means planning and designing your garden with regard to local conditions – taking into account the type of soil, the amount of rainfall, the temperature range,

and the prevailing winds. By choosing plants that naturally tolerate this given set of conditions, the plants will thrive better and less intervention will be required in the form of maintenance and control.

One example of working with local conditions is by adopting 'xeriscaping', an approach to gardening in very arid conditions which is increasing in popularity in the dry regions of the western and southern United States. Here, without xeriscaping, up to 50 per cent of domestic water consumption can go on garden irrigation.

Although the term itself is new, xeriscaping is based on the traditional methods of water conservation that have been practised for many generations by those living in the prairies. A range of strategies are employed for maximum water efficiency. These include improving the soil with organic matter to enhance water retention, grading, sloping or terracing the site to retain water, grouping plants with similar water needs, mulching, and above all, selecting native drought-tolerant species.

The following strategies should form the basis of eco-friendly gardening:

• Don't use chemicals. Instead, pick off larvae by hand and collect snails and slugs by using beer traps. By not using insecticides you also encourage pests' natural enemies, such as ladybirds that eat greenfly. Alternatively, practise companion planting, for example planting French marigolds alongside tomatoes. The roots of these marigolds exude a substance which deters whitefly.

• Compost organic waste matter from the kitchen, along with dead leaves and plant and grass trimmings.

• Improve your soil with compost and organic matter such as manure and keep it in good condition.

• Mulch to improve water retention and keep weeds at bay.

• Encourage birds, butterfiles, bees and other forms of wildlife by planting species that are attractive to them, such as nectar-producing flowers. Ponds will also attract a wide range of species.

• Set aside an area of the garden as a wild-flower meadow.

• Collect rainwater for garden irrigation. Avoid the use of sprinklers and hoses.

• Use hand tools whenever possible.

case studies

Eco design is not abstract, but firmly based in the existing conditions of a particular site or locality. The following section comprises case studies of individual eco homes from around the world, where a variety of environmentally friendly strategies have been integrated in response to very specific local conditions in terms of setting, climate and material resources.

'Thinking local' may be expressed in material use. It is hard to get more local than reusing timber from trees felled on the site, or earth excavated from the building plot. But designing with context in mind also means working with nature and climate: precision siting, angling of roofs and placement of openings to harness solar gain and natural ventilation. All of the houses featured here are in that sense indivisible from their particular contexts, whether that is a sloping wooded plot in Denmark, or a desert site in Arizona or the Australian bush.

At the same time, these houses offer much more than environmental worthiness. There is nothing 'bolt-on', for example, about the photovoltaics in Seth Stein's Finnish island retreat: the modules of the panels inform the building grid and infuse the design with purity and elegance. The quirky Wigglesworth-Till house, wrapped in straw bale and quilted fabric, playfully evokes its inner-city setting on the fringe of a main railway line. The Poole pavilions, each housing a different activity, challenge the whole notion of domestic space. All of the examples in this section, with their subtle attunement to natural rhythms, to light, sun, air and water, and to the surrounding landscape, provide richness of experience and delight.

In many ways, when it comes to eco design, we are on a learning curve. Many of the examples shown here have benefited from previous experiments with sustainable architecture on the part of their designers, with elements they have found to be successful refined and adapted in an ongoing process of fine-tuning. At the same time, many have been built in the teeth of scepticism and outright opposition from local authorities and planners who are uncertain how to apply building codes to eco design and construction methods. But however they have been achieved, the fifteen case studies presented here provide a good opportunity to show eco design in the round, where different systems and constructional techniques work together in a complementary fashion so that the sum is always greater than the parts.

STRAW-BALE HOUSE, LONDON, UK
ARCHITECT SARAH WIGGLESWORTH

OPPOSITE The L-shaped house with the library tower at the intersection of the two wings embraces a large garden. There is a railway line immediately adjacent to the plot.

OVERLEAF, LEFT The bedroom block at the end of the living quarters is insulated with a wall made of 550 straw bales.

OVERLEAF, RIGHT The office wing is covered with sheets of quilted silicon-faced fibreglass cloth that were made by a sailmaker.

A blend of high-tech and low-tech, the rustic and the urban, this sprawling L-shaped house occupies a plot right next to a main railway line in the heart of the London. The building comprises two wings and a library tower: one of the wings houses offices; the other, living space. Where the wings meet, there is a double-height space that serves as both dining room and conference area, bringing the live/work elements together. At the far end of the living wing is a bedroom block.

The office wing is raised up on 'gambions' – wire cages filled with lumps of recycled concrete. Springs between the gambions and the building help to absorb vibrations from passing trains. At the same time, the unusual wall cladding – cement-filled sandbags and quilted fibreglass cloth faced in silicon on the office wing, and 550 straw bales on the living wing – serves both as heat- and sound-insulation. The quilted fibreglass was made by a sailmaker. The straw bales are protected from the rain by corrugated sheeting, part of which is transparent to reveal the character of the construction.

Many of the materials, both internally and externally, are recycled. For example, the stairs leading up to the office wing are made of wood taken from trees that came down in the storms of 1989 and some of the windows in the house are framed with recycled railway sleepers found on the site. The wall insulation is mulched newspaper, while inside, the kitchen table, which also projects out onto the balcony, is of a material that looks like terrazzo but is in fact made of Ttura that uses recycled glass in place of marble chips.

Water conservation is another important part of the scheme's environmentally friendly credentials. Rainwater is collected and stored in two tanks and is used to flush the toilets beside the offices, in the washing machine and to irrigate grass and strawberries planted on the living roof. The solar-powered bathroom in the living wing has a composting toilet, which will eventually provide manure for the garden.

Another low-tech idea brought up to date is the larder, in this case built as a curved enclosure that separates the kitchen from the living room. Using cold air sucked up from the ground floor beneath the living area, it provides a cool enough environment to prevent perishable foodstuffs from spoiling.

For further information, see also

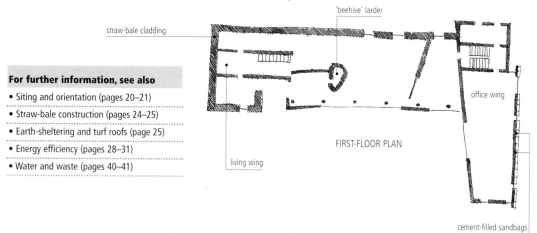

straw-bale cladding

'beehive' larder

office wing

FIRST-FLOOR PLAN

living wing

cement-filled sandbags

ABOVE The kitchen table is made from Ttura, a new terrazzo-like material created from recycled bottles.

LEFT The kitchen and living room are separated by a huge beehive-shaped structure which is actually an environmentally friendly fridge/larder. It works by sucking cold air up from the ground floor below the living area and keeps cool even perishables such as milk.

For further information, see also
• Siting and orientation (pages 20–21)
• Adobe and rammed earth (page 24)
• Using thermal mass (page 31)
• Natural and artificial light (pages 36–39)
• Concrete (pages 144–145)

CASE STUDY

OSBORN CLAASSEN HOUSE, TUCSON, ARIZONA, USA
ARCHITECT RICK JOY

OPPOSITE The site for the house was chosen for its drama as well as its seclusion. The architects have ensured that these elements are retained and that the environment is not damaged by the building.

OVERLEAF The gently warped 'butterfly' roof, made of weathered steel, provides protection for the rammed-earth walls as well as shade for the verandah.

Chosen for its drama as well as its seclusion, the site of this house is located in a small valley in the lower foothills of the Tucson Mountains, on the edge of the Saguaro National Forest. Surrounded by protected ridges, the local landscape is lush with desert flora and fauna, including numerous cacti.

An overriding aim of the scheme was to be as site-sensitive as possible: the design was developed after careful study of the existing vegetation, and the building was almost surgically inserted into the native desert so that no trees or cacti were destroyed during construction. Rainwater is diverted from the roof via a gutter that runs the full depth of the house but it is not collected for domestic use so as not to starve plants of water. On rare rainy days the water flows over the rocks in front of the verandah.

The rammed-earth construction of the house further emphasizes this site sensitivity and the architect's expressed intention to create a building that was rooted in the context and culture of its surroundings. The soil used to construct the house was not drawn from the site, but from three different sources in the immediate area – soils chosen for their colour and structural integrity. These were then slightly moistened, mixed with a small amount of iron-oxide pigment and 3 per cent Portland cement, and compacted into thick forms. The unreinforced exposed rammed-earth walls sit on concrete stem walls and spread footings. Some idea of the massiveness of such construction can be gleaned from the fact that the walls and foundations alone weigh 500 tonnes. The north and south walls are 0.6 metres thick and rise to nearly 5 metres. The centre valley of the roof slopes from 3.3 metres at the entry, to 2.5 metres at the massive fireplace core to the east, with its twin internal and external hearths. The softly warping 'butterfly' roof is made of weathered steel, extended out over the rammed-earth walls to serve as weather protection and to shield the verandah from the heat of the sun.

This very low-budget house is divided into two clearly defined rectangles, with living area, open kitchen, dining, pantry and bedroom on the south-east side, and guest room, bathroom and porch on the other. Interior walls, where they are not made of exposed rammed earth, are of painted drywall on steel studs, and the floors are of polished natural grey concrete. The windows and door frames are clear anodized aluminium and the windows are double-glazed. The bold modern design, high-quality craftsmanship and responsible use of building materials has created a house to stand the test of time.

RIGHT The guest room is located on the north side of the house. All the window and door frames are anodized aluminium.

BELOW The main living space of the house comprises an open eating, cooking and relaxing area. The rammed-earth construction of the external walls can clearly be seen.

ABOVE The floors are of polished concrete and the roof slopes down to the central fireplace core, which has a hearth both internally and outside on the verandah.

CASE STUDY

FLETCHER-PAGE HOUSE, KANGAROO VALLEY, AUSTRALIA
ARCHITECT GLEN MURCUTT & ASSOCIATES

ABOVE The south-facing side of the house is infilled with sliding panels. Four large corrugated tanks collect rainwater, from where it is diverted to a storage dam. All the water used in the house is collected rainwater.

OPPOSITE The north-facing (or sunny) side of the house has more minimal openings. Angled windows in the kitchen and living area are fitted with airflow panels.

In hot climates around the world, verandahs have traditionally served as intermediate spaces between indoors and out, shading the interior from strong sun and providing a place for informal living. A 'verandah house' which is essentially 'all-verandah', this contemporary version combines the flexibility and informality of open-plan living with a fine-tuned response to the basic elements of light, sun and air.

Siting the house involved careful study of both the ground and the local climatic conditions – contours, wind patterns, rainfall and the path of the sun. The minimal 150mm concrete slab that forms the foundation was positioned so that the house would be warmed by the sun in winter and cooled by through breezes during the hot months. The plan of the house is long and narrow, a sequences of spaces that ensured no area was isolated from its natural setting.

Exterior walls are made of timber on the outside and brick on the inside, with insulation in between. This serves to prevent heat from building up during the summer, but retains heat in the winter months. Underfloor heating (supplemented when necessary by the heat generated by a steel fireplace) provides space heating in winter. The electric heating cables rest on a 20mm layer of Styrofoam laid over the concrete foundation slab; on top is an 80mm concrete slab which forms the flooring throughout the house. Passive solar strategies, however, mean that very little additional heating is required.

The roofing is made of corrugated iron, a traditional building material of the Australian bush; the roof is angled to allow rainwater to run off into the four huge tanks that provide the house's water supply. Downspouts, detailed to prevent clogging from leaves, channel rainwater to the four corrugated iron tanks, from where it is funnelled to a storage dam below the house.

Folding, sliding, and angled openings offer maximum control of light, heat and air. On the south (shaded) side of the house are floor-to-ceiling panels, comprising insect screens on the inside, glass in the middle and wooden shutters on the outside. Angled windows in the kitchen and living area have airflow panels that can be opened during rainstorms to allow air to circulate but that prevent water getting in. Small sliding panels, with insect screens on the inside and slatted openings on the exterior, are strategically placed to deliver extra ventilation where required, such as behind the bedhead, for example. Daylight is regulated with electronically controlled power blinds that rise up behind the angled kitchen windows and the floor-to-ceiling windows in the living area.

Light, airy and open, the flexibility of the basic plan of this inventive house is matched by the thoughtful accommodation of a variety of weather patterns, harnessing the elements to provide both comfort and a sense of harmony with nature.

LEFT Sliding panels allow the interior to be fully open to the outdoors.

ABOVE Downspouts leading to the rainwater tanks are detailed to prevent clogging by eucalyptus leaves.

ABOVE The detailing of the entrance doors allows
the free flow of ventilation.

RIGHT The roof is made from corrugated iron, which is
a traditional bush building material. The angle
of the slope is designed to aid water runoff.

CASE STUDY
INTEGER HOUSE, GARSTON, UK
ARCHITECT COLE THOMPSON ASSOCIATES

ABOVE On the north-east side of the house, the roof is planted with sedum.

LEFT The large unheated conservatory faces south-west and makes the most of passive solar gain. Building components came from commercial glasshouse technology. There are photovoltaic panels and a solar collector on the roof. A spiral stair within the conservatory connects to an internal balcony off the main living area.

This demonstration house was designed to showcase constructional innovation, intelligent technology and environmental techniques, an integration of disciplines which gives it its name. No prototypes or experimental elements were used. Fast, cheap to build and with many up-to-the-minute 'smart' features, the house cuts CO_2 emissions by half and water use by a third.

Most of the elements were standard or prefabricated. Bathroom modules of the type pioneered for the off-shore oil industry were delivered complete with sanitaryware, plumbing, tiles, electrics, cupboards and finishes. The conservatory uses standard components from the glasshouse industry, and the ground beams, retaining wall and floor slabs were prefabricated.

The construction materials include sustainable, recycled and waste products, such as Western-red-cedar cladding and some timber flooring reclaimed from a demolished building. In addition, the house is highly insulated with recycled newsprint cellulose, is earth-sheltered to the north and all the paints and finishes used are organic.

Digital systems, controlled from a television monitor, were incorporated to enable fine-tuning of the internal environment. At the same time, the house was designed to be easy to update and upgrade. Skirting boards are removable to provide access to main cable routes, and a void behind plasterboard internal walls enables additional sockets or switches to be added at a future date. There's also a low-energy fridge, programmable hob, and automatic light sensors.

The large triple-height unheated conservatory dominates the south-west side. This glazed space works as a passive solar collector, with the concrete floor acting as a thermal flywheel. Automatic roller blinds reduce solar gain in hot weather. In addition, the house has a ground-source heat-pump system. This supplies warmed or chilled water to convector units, while its fuzzy-logic heat pump controls optimize energy use and allow temperatures to be set individually in each room. Domestic hot water is provided by a rooftop solar panel.

Cross-ventilation and stack-effect ventilation ducts in the kitchen and bathrooms provide natural ventilation, and there is supplementary ventilation by means of a fan powered by photovoltaic panels and wind turbine.

Water conservation is also key. Greywater is collected, filtered and treated for flushing the toilets. Taps have infrared and timed controls to minimize water use; aerated shower heads serve the same purpose. Rainwater supplies both hose pipes and a plant-watering system.

With its clever integration of high- and low-tech elements and its reliance on standard and prefabricated constructional components, Integer House is a house for the future that really works and sets new standards.

ABOVE The house is arranged so that the bedrooms are on the lowest level, with the kitchen, and living and dining areas on the first floor and a study and solar space on the topmost level.

RIGHT The solar space formed by the conservatory at the top level provides an extra living area. Automatic roller blinds offer shading when the sun is strong.

CASE STUDY

JOHNSON JONES HOUSE, PHOENIX, ARIZONA, USA
ARCHITECT JONES STUDIO

Known locally as 'The Dirt House', this contemporary reworking of rammed-earth construction occupies a half-hectare suburban site immediately adjacent to the 6,500 hectares of South Mountain Park, the largest municipal park in the United States. Despite the spectacular mountain views, no one had wanted to purchase the site because it was very close to an unsightly neighbourhood chlorinating tank. The resulting design responds to its location both by making use of local materials and by adopting cylindrical forms for outdoor areas to mask views of the tank and nearby houses. Metal fencing and open-block perimeter walls also act as semi-transparent screens between private land and the public park.

Oriented to take advantage of the beautiful views to the north, the house consists of two rammed-earth walls topped with a dynamic roof. The 0.6-metre-thick walls were made entirely from earth taken from the site: scoops of dirt were slightly moistened, mixed with 3 per cent Portland cement, and loaded into plywood forms. Then a handheld pneumatic tamper compacted the dirt into a rock-hard 150mm layer and the process was repeated until the desired height was achieved and the forms could be removed. The walls are left nude. Because heat conducts through compacted dirt at a rate of only 25mm per hour, the interior surface of the walls remains at a constant room temperature throughout the year.

Other construction materials used in the house either have significant recycled content or are resource-efficient. These include plywood bonded with toxin-free adhesives, engineered composite framing timber, rusted-steel wall cladding and concrete block manufactured with fly ash. The ceiling is made of vertical-grain Douglas fir. Very few surfaces are painted, but where they are, finishes are low- to no-VOC paints.

Rainwater collected from the roof is used to irrigate the indigenous drought-resistant landscaping. The roof slopes to a large scupper projecting from the round central turret that houses the internal staircase. Stainless-steel chainlink hung from rusted black iron plates directs the water to a 5.5-metre holding area. When this holding area fills, carefully placed overflows allow subsequent distribution to the landscaping.

A central corridor forms the main axis of the plan, with a double-height open kitchen/dining/living area to the north, leading out to a patio and pool, and more private areas, including bedrooms and an office, located on the south side. An internal window connects the kitchen with the adjacent game room. The double-height living area is dominated by a glass wall, dramatically cross-braced with steel. Elliptical stairs in the central stair turret connect to an upper level, where the master bedroom and a studio are situated. Exterior 'rooms' located on all sides of the house can be comfortably used all year round.

ABOVE The rammed-earth walls of the house were constructed of dirt taken from the site, which is dramatically located on the fringes of a spectacular park.

Every habitable area enjoys natural daylight and natural ventilation. Extensive overhangs and a rusted steel sunshade keep off strong summer sun yet allow winter sun to penetrate. All doors are weather-stripped and the roof is insulated with blown cellulose. The large north-facing glass wall floods the main living space with soft, natural daylight; the glazing is high-performance and is further shielded from direct heat gain by the extended rammed-earth walls.

On the upper level, the central corridor connecting the master bedroom and the studio consists of a glass bridge, composed of 25mm-thick clear glass panes between wooden glulam beams. Directly overhead and running the full length of the corridor is a skylight. The result is that natural light penetrates all the way down to ground-level spaces. In the summer, the skylight is protected with a 'green house' fabric to deflect heat but allow light through.

Throughout the house, most of the electric light fixtures are state-of-the-art high-efficiency fittings. A lighting control system allows maximum dimming and monitors electric consumption. Along with the high degree of insulation provided by the rammed-earth construction, high-performance glazing and super-insulated roof, little additional energy is required to condition the space, and utility bills are consequently low.

The rugged earthen construction, use of sandblasted concrete block, the expressive natural interior surfaces of black granite, glass and Douglas fir, and the recurring use of circular, cylindrical forms, create a house that is not only eco-friendly and energy-efficient but one that is at ease in its setting. Appropriately, embedded in a living-room wall is a stone from the chimney of Shiprock, the first Arizona home designed by Frank Lloyd Wright.

ABOVE Rainwater is diverted from the roof to a scupper projecting from the central staircase tower. 'Rain chains' direct the water to a holding area, from where it is distributed to irrigate the landscape.

OVERLEAF The deep overhang of the roof prevents the interior from overheating; extensive high-performance glazing provides maximum natural light.

ABOVE An adjustable steel sunshade deflects the heat of the summer sun, but allows light through in the winter months.

LEFT The rugged rammed-earth wall is left bare, providing a textural backdrop to the main living area with its north-facing window wall.

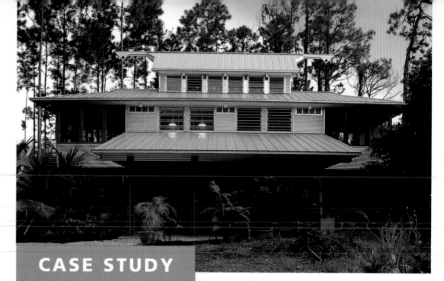

CASE STUDY

PALMETTO HOUSE, MIAMI, FLORIDA, USA
ARCHITECT JERSEY DEVIL

Located on the fringes of the Everglades in southern Florida, a part of the world where levels of heat and humidity are often high, the design of Palmetto House proves that by adopting local vernacular traditions, along with other sophisticated strategies for passive ventilation, comfortable living conditions can be created without recourse to air conditioning. The house is on three levels: workshop on the ground floor, living space on the first floor, and a top-floor loft.

For generations, local farmers built their houses with deep overhangs to shade walls, porches at either end and unobstructed interior space to allow cooling air to flow through. Similar features can be seen at Palmetto House. The plan is in the form of a cross, with the longer sides of the building aligned north/south and the living space aligned east/west and raised up a level to catch the cooling south-easterly breezes. On the south side, deep eaves keep the sun off exterior walls in the summer; awnings shade the windows so they can be kept open, even during storms. At either end, screened porches shield interior spaces from the sun; inside, there are few partitions or walls so that air flows freely – the main living space is an open-plan kitchen/living room/dining room. The floor of the loft is metal grating, which also allows air to circulate and permits light to filter through, and the walls are lined in louvred windows. Lush subtropical undergrowth shades the house at the lower level.

Material use also had a role to play in natural cooling. The exterior walls and roof are clad in corrugated aluminium, which reflects the light and heat; high-mass concrete, used in the construction of the ground floor, keeps this part of the house cool. The upper floors are lightweight timber-frame construction.

To supplement these natural cooling features, walls are also vented. Radiant barriers made of high-emissivity metal foil inside the walls and roof trap heat before it has a chance to reach the interior and expel it through high-level vents, drawing cooler air in from the shaded eaves. A dozen mechanical ceiling fans keep air moving internally.

Almost all of the house's hot water is heated by solar panels installed on the portion of the roof which faces south, and the water is circulated by a pump powered by photovoltaic cells. There is also a greywater system, with flow-control taps to conserve water.

Unlike air-conditioned homes, which must be hermetically sealed, this house is open to the elements and to the sights, sounds and smells of the natural landscape. Weather permitting, the back porch doubles up as an additional bedroom.

ABOVE Deep overhanging eaves, screened porches and high-level wall vents cool and ventilate the interior without air conditioning. The site features lush subtropical vegetation.

OPPOSITE The loft space on the top level is floored with metal grating to allow air and light through. The walls are lined with louvred windows. Plexiglass on the ceiling diffuses fluorescent light. No supplementary lighting is needed during the daytime in any part of the house.

DETAIL OF VENTED WALL

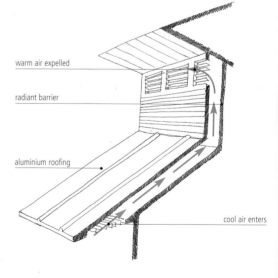

warm air expelled

radiant barrier

aluminium roofing

cool air enters

ARAUCARIA, POMONA, QUEENSLAND, AUSTRALIA
ARCHITECT GABRIEL & ELIZABETH POOLE DESIGN COMPANY

Taking its name from the Araucaria family of pines, which includes the hoop, bunya and Norfolk Island pines native to this part of Queensland, this small one-bedroom house perches lightly on a gently sloping site with views of the Pacific Ocean. As the remote site has no connections to water, sewage or electricity, it was important both to design for maximum self-sufficiency and to come up with a structure that could be readily assembled using prefabricated elements. The cottage and storage shed were built in three months, and all the timber construction was carried out by a single carpenter. The house could easily be extended at a later date.

Before coming up with the design, the architects made an intensive analysis of the site to determine optimum positioning, looking in particular at the local topography, the prevailing winds and the pattern of rainfall. One key consideration was wind strength: the area has a high-wind classification, due to the threat of cyclones.

Accordingly, the structure is a lightweight simple rectangular box, thoroughly braced and with a low-pitched roof to minimize wind resistance, the design recalling local vernacular farm buildings with their verandahs and broad overhangs. Maintenance is minimal. Rainwater is collected and stored in tanks further up the slope. A combination of bottled gas and solar power is used to heat water and run the refrigerator and stove. Supplementary heating is provided by a slow-combustion burner positioned in the centre of the house.

Because the site is so exposed, it was important to balance lightness and openness with parts of the house that are more enclosed and private. While expansive glass windows draw in the views, the covered verandah/entry point on the north side (the sun side) provides a more sheltered aspect, and sliding internal panels allow spaces to be left open or shut off. Adjustable glass slats in the floor-to-ceiling windows allow cross-ventilation even when it is windy.

The house is very compact, measuring only 8 metres by 10 metres, and a modular plan was adopted to fit in all the necessary accommodation for comfortable living. Divided into six modules of 3.3 metres by 4 metres, the layout comprises kitchen, verandah, bedroom and study at one module each, a bathroom at half a module and a main living area of one and half modules. Built-in furniture, storage, shelving, cabinets and window/seat guest bed make the most of tight planning.

Light, simple materials are used throughout: hardwood (ironbark) posts, prefabricated timber trusses and beams for the basic timber frame, steel for bracing and for the roof, along with plywood, cedarwood louvres, corrugated steel and acrylic for external walls. Interior finishes include plywood for ceilings, hoop-pine battened sliding panels, floorboards made of spotted gum, and turpentine decking boards. Although the materials are basic, they offer depth of visual interest and tactility, with the warm tones of the wood sharpened by black-painted structural elements and set against the silver of corrugated sheeting.

The site was formerly part of a dairy farm, and is now overgrown with wattle and groundsel. A long-term aim of the owners is to replant the lower portion of the land with the type of rainforest species that once used to thrive here.

LEFT The design of this compact, self-sufficient one-bedroom home echoes the form of local farm buildings. Much of the construction was prefabricated off-site.

For further information, see also

LEFT The structure is thoroughly braced to resist high winds. The low-pitched roof is made of steel, and overhangs provide sun screening.

ABOVE The study is on the south side of the cottage. Adjustable glass slats provide cross-ventilation to cool the interior, and a window seat doubles up as guest bed.

ABOVE The layout is modular, with the kitchen, verandah, bedroom and study at one module each, and the living area at one and a half modules. A slow-combustion burner in the centre of the open space provides additional heating in the winter.

RIGHT Sliding partitions provide flexibility. The slatted panels allow air to circulate.

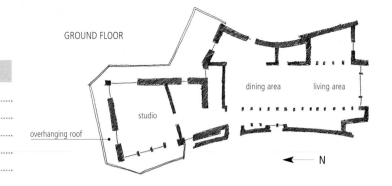

GROUND FLOOR

dining area   living area

studio

overhanging roof

N

RIGHT The hanging box of the studio, with its perimeter balcony, is jettied out over the sloping site. The site gradient drops 8 metres overall.

LEFT External walls clad in cedar shingles twist round the existing mature beech trees to create an undulating form.

CASE STUDY
AARHUS, DENMARK
ARCHITECT CHRIS THURLBOURNE

Enshrining many passive eco strategies, this low-budget house occupies a difficult sloping site in a mature beech forest. Because the land drops 8 metres, the design had to be conceived almost sculpturally to fit the necessary accommodation onto the site. Six beech trees were felled to clear the site but all the wood was re-used in the construction.

It is a highly crafted house and one which reflects an interest in natural materials. The exterior walls, which twist outwards like sails, are clad in cedar shingles and narrow horizontal pine boards. In the skylit central triple-height room, individual spaces are delineated both by subtle changes in level and a series of 21 handmade mahogany columns that form an internal screen. These columns comprise two slices of mahogany strips 6 metres long, laminated and bolted on each side to a strip of plywood. The colonnade serves to separate the main living spaces from ancillary areas, including stairs, guest bathroom, corridors and kitchen. A step down from the entrance-cum-dining area is the living area. Here concrete flooring laid over underfloor heating gives way to wooden floors to underscore the change of emphasis.

A bridge across the central atrium at first-floor level connects a bedroom at one end with a library and bathroom at the other. On the opposite side of the house to the main bedroom are two self-contained areas, one a second bedroom and the other a studio, which jetty out over the sloping site. From the first floor there is access to two roof terraces.

The form of the house responds not only to the site, but also the conditions of natural light. In the winter months, south-facing and roof windows allow light to penetrate the entire house and hit the north wall. In the summer, the existing trees create a 'blanket' of shade that prevents the house from overheating. The construction is well insulated, and Velfax windows provide additional energy savings. There is only one small window on the north-facing side (where there is also a nearby apartment block), which also helps to provide a sense of privacy and enclosure.

Like most of the homes in this part of Denmark, this one is connected to the district heating system. At the central district plant, 'waste' energy from the production of electricity from coal is used to heat hot water which, in turn, is pumped to homes in the district. This means that houses do not need individual boilers. In this home, radiators line the walls at second-floor level and help to keep the temperature constant throughout the lofty 8-metre-high space.

ABOVE The handcrafted mahogany and beech colonnade serves as a transparent spatial divider, separating the main living areas from the stairs and circulation space. The central portion of the house is virtually triple height.

RIGHT The living area is set down a level from the dining area and is further defined by a change in flooring. Light pours into the space through large south-facing windows.

LEFT The central atrium or glasshouse connects the two wings of the house and is planted with exotic hot-country plants. The soil used to construct the rammed-earth walls was taken from the site.

SECTION THROUGH RAMMED-EARTH WALL

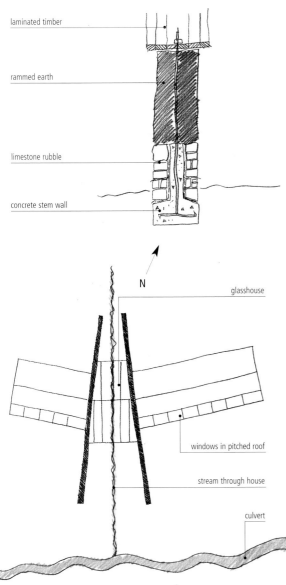

laminated timber

rammed earth

limestone rubble

concrete stem wall

N

glasshouse

windows in pitched roof

stream through house

culvert

CASE STUDY
WEST LAKE BRAKE, PLYMOUTH, UK
ARCHITECT DAVID SHEPPARD

Located on the south-west coast of England, this seaside house encloses a mini-Eden project of its own, in the form of an integral glasshouse planted with a variety of species native to Mexico, Spain and Australia. Equally exotic is the rammed-earth construction of the principal walls of the house, a traditional building method more common in hot arid regions.

Soil samples from the site were tested and analysed at the University of Plymouth to determine if the clay was of a suitable consistency. A less scientific assessment was done on site; this entailed rolling some clay into a ball about the size of a tennis ball and dropping it from waist height. The rule of thumb is that if the clay ball breaks into four pieces, as it did, it is the right consistency. No binding agent was used; compacting the clay from 1.8 cubic metres down to 1 cubic metre forces out all the air and water and makes the material strong and rigid.

The two main rammed-earth walls – 500mm thick, 22 metres long and an average of 2.4 metres high – form a V-shape from which the two wings of the house branch. In rammed-earth construction, particular care must be taken to prevent moisture penetration, especially from above and below, so on the lower level of this house, a concrete stem wall is faced with random rubble limestone set in earth joints to prevent rising damp. Above, the rammed-earth walls are protected by a 650mm-wide capping of laminated timber which allows water to drip off well clear of the walls.

In an area where high winds are not uncommon, there was also the need to safeguard against wind loading and wind erosion. Steel reinforcing bars running up through the concrete stem wall and into the rammed earth are tensioned top and bottom to provide additional stability. The upper levels of the house have a timber-frame structure made of green oak. Again, the timber is tensioned with yachting wires, a nautical reference that suits the area's naval links.

The house is in two parts, with a bridge across the conservatory at first-floor level linking the kitchen and dining area with the main living space. All the bedrooms are on the lower level, where the thick walls provide privacy and soundproofing. During the day, the glazed conservatory heats air which rises and warms the first-floor living areas. Windows along the eaves provide stunning views of the sea and can be opened in the summer months on the sea side of the house to provide natural ventilation and cooling. The roof itself is made of reclaimed natural slate and is highly insulated; steep and aerodynamically pitched, it minimizes wind resistance and helps to draw cool coastal breezes indoors in the summer months. Interior finishes are simple: for instance, flooring is of basic plywood sheets varnished to a high shine.

One of the more surprising features of the house is the stream that runs through it. A water diviner pinpointed the location of an existing culvert, part of an underground water system for irrigating local pastures, and water was diverted to run through the building in a channel which loops around to rejoin the main stream further down the hillside away from the house. The channel or trough is made of reclaimed limestone cut in a deep V. The water, which helps to irrigate the plants in the mini-Eden, also serves to cool the building by up to 3° centigrade.

ABOVE The upper structure is timber frame, made of green oak and tensioned with yachting wires.

RIGHT Windows along the eaves flood the living area on the first floor made from plywood sheeting.

OVERLEAF, LEFT A bridge across the conservatory at first-floor level connects the two wings of the house.

OVERLEAF, RIGHT The two massive rammed-earth walls are faced with rubble limestone at the lower level and capped with laminated timber to prevent moisture penetration.

For further information, see also

• Timber and timber-frame (pages 22–24)

• Energy efficiency (pages 28–31)

• Wood (pages 126–133)

• Paints, varnishes and seals (pages 156–159)

CASE STUDY

VILLA VISTET, SWEDEN
ARCHITECT LANDSTRÖM ARKITEKTER

ABOVE This timber house, originally built as an exhibition prototype, can be readily taken apart and reassembled.

RIGHT All surfaces and finishes are entirely natural and non-toxic, enhancing the beauty of the wood.

Timber construction is traditional in many parts of the world. In northern Europe, particularly Scandinavia, where there are extensive forested areas, timber is the cheapest, most readily available and most local of all building materials, which makes it an environmentally friendly choice. The nature of wood, and the typical dimensions of timber members, also results in simple, well-proportioned buildings.

The idea behind the design of Villa Vistet was to develop a modern log house with a floor plan and services to suit a contemporary lifestyle. Originally built as an exhibition house, the entire structure is planned like a construction kit, and can be taken apart and reassembled in another location. Similarly, the design allows flexibility in terms of technology, offering the opportunity to update and renew services to the latest standards. For example, although solar panels were not installed on the exhibition house, they could easily be sited on either walls or roof. Similarly, depending on where the house was located, a heat pump taking energy from water or the ground could be installed. Rainwater collection and other water-saving features such as composting toilets could also be included.

Renewable, recyclable, locally sourced wood accounts for 90 per cent of the materials used at Villa Vistet. The timber walls are 200mm thick. These generate a good interior climate by assimilating heat and humidity and they also promote natural cooling and ventilation. The main entrance, staircase, storage area and toilet are located on the north side of the house, where small windows retain the heat, while the kitchen, living room and bedrooms face south and have larger windows to benefit from solar gain. The roof and floor framework are insulated with wood shavings but other recycled materials, such as cellulose, could also be used. Heat levels can be controlled in every room.

All the internal and external finishes used are non-toxic. The floor was treated with vegetable oil, while the interior walls are painted with egg tempura. Window frames are made of heartwood, which does not need potentially hazardous impregnation, while the external walls are painted with lime-based paint, a traditional organic finish.

CASE STUDY

THE BOUNDARY HOUSE, TUNBRIDGE WELLS, KENT, UK
ARCHITECT WINTER & MONK

ABOVE The wooded site of the house is right next to a cricket ground and borders a main railway line to the rear.

RIGHT The undulating form of the building's footprint was designed to minimise disruption to the wooded site. The minimal concrete-pad foundations required by the timber-frame construction were fitted among the roots of existing trees.

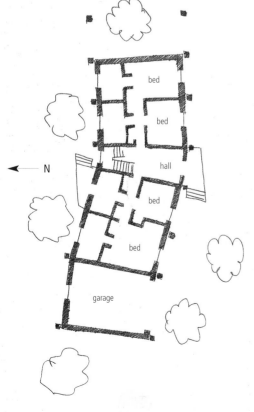

GROUND FLOOR

bed

bed

hall

← N

bed

bed

garage

Proof that eco design does not mean unattractive, this sophisticated contemporary family house has won three major awards. At 302 square metres it is spacious, with four bedrooms, three bathrooms and a utility room on the ground floor, a mostly open-plan first floor of living, dining and kitchen areas, plus a separate studio, a study and a raised deck.

The narrow wooded site is right next to a cricket ground – hence the name – and there is a railway line to the rear. The house was built in a clearing, its gently undulating plan designed to avoid the trees (only two had to be felled). The timber-frame structure also meant that there was no need for conventional foundations, which would have destroyed more trees. Instead the frame is raised off the ground by glulam timber columns. These rest on individual concrete-pad foundations carefully inserted among the roots of the existing oaks.

The structure is a development of the traditional post-and-beam timber frame but uses glulam beams, posts and trusses fixed with steel bolts. In between the glulam are beams of laminated plywood, which are as strong as solid studs but contain less timber. The 'breathing-wall' method of construction was adopted, with the exterior clad in boards with an organic stain and the wall cavity filled with blown-cellulose insulation. The copper roof is also highly insulated and will eventually weather to blend with the woodland. The north section of the roof is raised slightly to introduce a clerestorey and bring more natural light into the interior.

The whole house makes the most of solar gain. The upside-down layout means the living areas at the top benefit from rising heat (and views); service areas are to the north; bedrooms and living areas face south. On the north side, there is also a larder, with external vents for natural refrigeration. Windows on the north are kept to a minimum, which also reduces the noise of passing trains, while the south side incorporates extensive glazing. All windows are high-performance, low-E coated triple-glazed and argon-filled, in Scandinavian softwood frames.

The proximity of the railway line meant that a mechanical ventilation system was adopted to reduce noise. Fans bring fresh air in and expel smells and condensation, while the ventilation system is connected to a heat pump and heat exchanger, so incoming air is warmed by the waste heat generated by appliances in the kitchen and hot towel rails in the bathroom. Solar panels on the roof provide energy to heat water. On very cold days, supplementary space heating is provided by a wood-burning stove, in the centre of the main living area.

Rainwater collected from the roof and channelled down copper pipes is stored in an underground tank beneath the garage where it is purified for household use. Drinking water, however, comes from the mains supply.

ABOVE A bridged landing at first-floor level connects the main living spaces, which are warmed by rising air. The bedrooms are located on the ground floor.

RIGHT An extensive terrace at the upper level provides outdoor living space among the trees. The roof is covered with copper, which will eventually weather to green, blending in with the woodland.

OVERLEAF The first floor is largely open-plan, made possible by the use of glulam timber members, which allow longer spans. A wood-burning stove, located in the centre of the living area, provides supplementary heating.

For further information, see also

• Siting and orientation (pages 20–21)

• Timber and timber-frame (pages 22–24)

• Ventilation (pages 26–27)

• Energy efficiency (pages 28–31)

• Wood (pages 126–133)

CASE STUDY
PRIOR HOUSE, AVALON, AUSTRALIA
ARCHITECT STUTCHBURY & PAPE

New-build projects offer the greatest scope for thorough-going eco design. But, as this renovation and extension of a 1950s cottage demonstrates, additions and internal rearrangement can go a long way towards enhancing the environmental credentials of an existing property. Here, the basic brief was not simply to provide more space, but to connect the house more fundamentally with its setting, working with the site and the local climate to produce a design that was in tune with nature.

The success of this approach can be gauged from the fact that although local temperatures range from 9–11° centigrade in winter to 30–32° centigrade in summer, the temperature variation in the house is only 3 degrees, a stability achieved by using thermal mass and natural ventilation. On a deeper level, working with the climate has resulted in a house which feels more natural, open to the play of natural light, cooling breezes and the warmth of the sun.

The form of the extension emerged from a desire to connect the house with both the garden and the courtyard on the upper level. Existing stairs up to the garden from the back of the house were enclosed and a new mid-level introduced, which opens out onto the courtyard on the north side, and, via a new indoor/outdoor bathroom, to a private deck on the south side. The stairs are then turned back over the top of the house to connect with a new top-level bedroom with views through the surrounding trees to the ocean.

The north side of the new extension is open, allowing winter sun to penetrate the interior and cool summer breezes to flow through. The light, indoor/outdoor quality is matched by an effective use of simple materials: recycled blackbutt timber for the main framing and for tongue-and-groove flooring on the upper level; western-red-cedar-framed doors and windows; and hoop-pine plywood cladding for the ceilings. High-quality craftsmanship and attention to detailing were important elements that contributed to the success of the result.

Working with the pre-existing levels and the climate has resulted in a house that offers much richer spatial experiences, from the cool contemplative haven of the top-level bedroom with its soothing views, to the shower room that connects directly to an outdoor decked area, to the wooden walkways that provide a sequence of interesting transitions from place to place and level to level.

ABOVE This clever addition to a single-storey cottage in Australia boasts a corrugated-iron roof that emphasizes the slope of the land. A wooden deck connects house and garden, and large sliding doors provide natural ventilation.

LEFT A short flight of stairs provides access to the living room which has a bed alcove built in against the north window wall.

ABOVE The extension provided the opportunity to improve the connection between the house and its surroundings, as evidenced by this indoor/outdoor bathroom.

RIGHT Existing stairs up to the garden from the back of the house were enclosed. Window and door frames are made of western red cedar.

For further information, see also

• Siting and orientation (pages 20–21)
• Timber and timber-frame (pages 22–24)
• Natural and artificial light (pages 36–39)
• Wood (pages 126–133)
• Glass (pages 146–147)

CASE STUDY

PINE FOREST CABIN
METHOW VALLEY, WASHINGTON, USA
ARCHITECT CUTLER ANDERSON

ABOVE The cabin is sited on a slope and sits on concrete piers to minimize disruption to its setting. Openings are more minimal on the exposed north-facing side.

LEFT The south-facing side of the cabin is virtually all glass to benefit from passive solar gain and bring views and light into the interior. Heat loss during the winter months is prevented by double-glazing with low-E glass.

Situated on a gently sloping wooded site in Methow Valley, Washington State, this contemporary cabin serves as a year-round retreat, with substantial covered outdoor areas to provide both shade from the sun in summer and shelter from the snow in winter. The lightweight timber construction blends perfectly with its semi-arid pine forest setting.

The harmony achieved between house and context arose out of the architect's determination to bring out the salient qualities of the setting in the design of the building's form – in other words, to make both fit together as closely as possible. After an extensive survey of the topography of the site, it was decided to make the structure vertical like the trees.

The cabin is extremely site-sensitive. No trees were chopped down during construction and the simple framed box of the house sits up on 14 pyramid-shaped concrete piers, so that the house almost appears to hover over its site. Because timber-frame construction is substantially lighter than other forms, notably solid masonry, these foundations could be relatively minimal. The concrete piers interrupt the site as little as possible and accommodate the sloping gradient in an evocative fashion.

The house itself is 5 metres wide; decked and terraced areas at both ends are sheltered by the overhanging roof. Extensive glass on the south-facing side maximizes natural light in the interior. Low-e double glazing helps to insulate the interior, preventing excessive heat loss during the cold winter months.

The plan is compact, with a simple entrance on the ground floor leading to a living/dining/kitchen area, and two bedrooms and a bathroom upstairs. One of the bedrooms is fully enclosed; the other is on a mezzanine level with expansive views over the surrounding countryside. A small powder room is under the stairs.

Reflecting the architect's respect for natural materials, the locally sourced timber constructional elements and framing members are boldly expressed, like a kit of parts. Even where there are partitions, the lower portion of the wall is left uncovered to reveal the feet of the studwork.

This honesty of construction, whereby the building reveals its means of assembly, enhances both its natural quality and the fit between house and landscape.

ABOVE Interior surfaces and finishes are very simple, with the timber structure honestly exposed for a sense of rhythm and detail.

RIGHT The cabin has extensive terraces at front and rear, both of which are covered with deep overhangs to keep sun off in summer and snow off in winter.

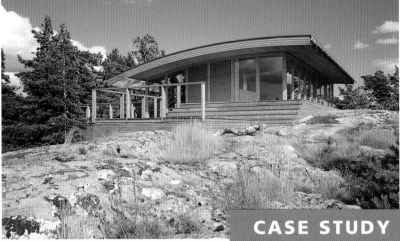

For further information, see also

• Siting and orientation (pages 20–21)

• Timber and timber-frame (pages 22–24)

• Energy efficiency (pages 28–31)

• Heat and power (pages 32–35)

• Water and waste (pages 40–41)

• Wood (pages 126–133)

CASE STUDY
ARCHIPELAGO HOUSE, FINLAND
ARCHITECT SETH STEIN

Few settings are more remote or more dramatic than this small uninhabited island in the Finnish archipelago, some 80 kilometres west of Helsinki. Situated in the centre of the island and on its highest point, 15 metres above sea level, this solar-powered summerhouse and yoga retreat sits on a wooded granite outcrop with expansive views over the sea towards Estonia. It is the perfect location for relaxation and contemplation.

Solar technology is literally built into the design: the building grid is based on the dimensions of a standard photovoltaic panel (1.35 metres by 3.3 metres). Twelve photovoltaic units are located on the roof edge, which corresponds in depth to the curved laminated timber beams that span 8 metres internally. These units generate enough power to pump and heat sweet water from a well sunk below the island and brackish water from the sea, which is used for washing and flushing the toilet. Solar energy also supplies all the other power requirements in the house – including an electric toilet – and systems such as heating and the refrigerator can be activated remotely by mobile phone.

Because of the remoteness of the site, the structure was prefabricated in local workshops during the winter months. It was subsequently transported across the archipelago, carried up to the site and assembled manually over two summer seasons. The materials used include local timber from a managed plantation, glass and local stone. There is also low-e lighting and all the indoor finishes are non-toxic.

Enclosing an area of 110 square metres, the summerhouse is simply planned, with a large south-facing living area giving out onto the perimeter decking. The kitchen island is placed at the core of the living area, with a single bedroom on the east side partitioned by a sliding screen made of woven paper yarn. In the bathroom, a washbasin has been fashioned from a stone found on the island's shore and is positioned to face the rising sun. Outside, a plunge pool is set into the deck, its contours echoing the form of an Alvar Aalto vase, a design that is emblematic of Finland.

With the gently curving roof following the line of the rocky outcrop, and the effortless integration of structure and solar technology, the design of Archipelago House achieves an elegant simplicity which is ideal for its function as a place of retreat.

ABOVE Because the site is inhospitable during the winter and is also so remote, the main structure had to be prefabricated on the mainland and transported for assembly over two summer seasons.

LEFT The serenity of the location and the elegance of the design make for a perfect retreat. The house is only used during the summer months.

ABOVE A contemplative bathroom includes a washbasin fashioned from a stone found on the island's shore.

LEFT Decking all the way round the house extends the living space outdoors. There is a plunge pool inset in the deck; its contours echo the form of an Alvar Aalto vase.

ABOVE The large living space has the kitchen at its core. All the glass is triple-glazed and the timber comes from local sources.

OPPOSITE Located on a small uninhabited island in Finland, the house has twelve standard photovoltaic panels located on the edge of the roof. These determined the building grid and provide all the electricity the property needs.

CASE STUDY

POOLE HOUSE, LAKE WEYBA, QUEENSLAND, AUSTRALIA
ARCHITECT CLARE DESIGN

OPPOSITE Three individual structures or pavilions connected by wooden walkways comprise the 'house'. These light structures are raised over the ground to avoid damaging the rare grass that grows here.

A house that is actually a row of houses connected by wooden walkways, the three structures that comprise the Poole House appear to float over the ground, an impression reinforced by the light, airy, almost transparent construction. Extremely site-sensitive, the individual structures hang 60cm above the ground, in order to avoid damaging the rare wild wallumgrass that grows all over the site.

In a hot, dry climate it is no particular hardship to have to venture outdoors if you want to move from the kitchen to the bedroom, for example. Accordingly, each of the three units serves a different function. One of the buildings is the bedroom; the middle structure is the 'wet room', with shower, bath and toilet; while the main 'house' accommodates everything else, from cooking and eating, to working and relaxing, within 170 square metres of space. The entire structure is 35 metres long. A particular feature of the design is the way that many elements that would normally be 'built-in' are actually built out, borrowing space from outside the basic shell of the units. These include cupboards, seats, hearth, shower cubicle and other alcove-like spaces which hang like rucksacks outside the rooms.

Another architectural innovation is the double-hanging construction of the roof, which serves both to cool the interior and allow maximum natural light through. Awnings on the outside are stretched across an inner layer of PVC sheeting, with the space between the two serving to cool the air. Lightweight steel poles support the roof structure and rest on metal floor beams, while the main structure consists of light hollow-steel sections set 2.4 metres apart and infilled with wall panels, windows and doors.

The external walls are made of smooth galvanized iron, a metal chosen because it oxidizes to a dark grey that blends in with the natural surroundings. All doors and windows are framed in heavy local softwood that is painted to reflect the colours of the trees, flowers and other vegetation. The decking is made from local hardwood, while the interior flooring is pine plywood. In the shower cubicle, the floor is made of a few square metres of metal industrial flooring, which obviates the need for a drain.

Appropriately for a house situated in a dry area, water conservation is a crucial part of the design. The household relies for all its water needs solely on rainwater collected from the roofs and stored in tanks. Recycled greywater is used to irrigate the garden, and there is a composting toilet.

Natural ventilation is also important. The individual structures are sited and designed to take advantage of cross-ventilation and there are ingenious vents that can be adjusted like blinds. The verandah on the north-facing side of the main house is shaded with tensioned PVC fabric awnings, while awnings also cover the deck areas that connect the individual pavilions.

For further information, see also

LEFT A particular design innovation is the way space is borrowed from outside the basic shell of the structures, to create 'built-out' features such as this bed alcove.

RIGHT The bed alcove viewed from the exterior, where it protrudes from the main external wall. The walls are made of galvanized iron.

LEFT The household relies solely on rainwater collected from the roofs and stored in cylinders set between each of the three pavilions.

ABOVE The outdoor decking is made of local hardwood. Lightweight steel poles support the roof structure; tensioned awnings shade the north-facing terrace.

RIGHT An internal view of the 'living' pavilion which accommodates cooking, eating and relaxing. The bedroom is in a separate pavilion at the opposite end, while the central pavilion houses the wet areas of bathroom and shower.

SURFACES & FINISHES

Buildings consume vast quantities of materials. It takes half a hectare of forest to construct an average home in the United States, while the manufacture of its concrete foundations generates some 9,000 kilograms of carbon dioxide emissions. As well as the impact on the environment of the constructional materials, about which individuals may have little choice, there is also the impact of the materials employed as surfaces and finishes to consider.

Selecting environmentally friendly materials and using them efficiently are central to eco design. Often it is simply a question of substitution: choosing flooring made of bamboo, for example, over one made of an endangered hardwood. Defining what makes a material 'green', however, involves many complex issues to do not only with the sustainability of natural resources, but also with the energy consumed in their processing and transportation and their durability, as well as their impact on human health. It is not always possible to avoid non-green materials, but with care they can be used minimally and in such a way as to enhance the environmental friendliness of the design.

When it comes to evaluating the environmental impact of a material its 'embodied energy' is important. This is the sum of the energy required at all stages of production: the extraction of the raw material, its transportation to a factory, the energy used in processing, the transportation of the processed material to the point of use and the energy used during construction. Obviously, the fewer steps between origin and use, the lower the embodied energy and the less waste. Very highly processed materials such as metal and plastic have high embodied energy while local materials that do not have to travel far between origin and use have lower embodied energy.

Equally significant is the way that a material is used. Using small quantities of a material that is high in embodied energy may be beneficial if this improves overall durability or structural performance. The longer a building or element of a building lasts, the lower its

RIGHT Metal is a material that is high in embodied energy but in the right circumstances or application can be an effective eco choice because of its durability and strength. This shower stall clad in corrugated sheet in an Australian house makes use of a material widely used in local vernacular buildings.

environmental impact. Similarly, it may be justified to use materials such as concrete or brick, which have relatively high embodied energy, because their high thermal mass significantly reduces a building's energy needs over its lifetime.

Another factor is the recycled content of a material. Direct recycling – using materials salvaged from other buildings saves a great deal of energy and protects natural resources from depletion. But the issue becomes more complex in the case of producing materials from consumer or industrial waste, such as plastic sheet made of recycled vending cups. Where that waste would have ended up on a landfill site, which is the case for most consumer waste, recycling is generally regarded as preferable, despite the fact that the recycling process itself consumes energy. Closed-loop recycling, for instance recycling carpet as carpet or toothbrushes as toothbrushes, is also better than recycling that produces a material or product of lower grade.

The availability of green materials and products is improving all the time and will continue to improve with consumer demand (a study conducted in 1990 found that 80 per cent of the public would prefer to purchase green products if given the option). At the same time, it is important not to take the claims of manufacturers and producers at face value. Materials or products that are certified as environmentally friendly by independent bodies are more trustworthy than those which may simply be 'greenwashed' by manufacturers or producers eager to retain their market share.

OPPOSITE An exposed stone wall in a Shropshire farmhouse has a rugged natural honesty. The bedcover is made of felted wool.

Wood

Wood is an extremely versatile material, with a huge range of uses in construction and finishing. From an ecological point of view, one of its strongest credentials is that it derives from a renewable, living source, and one that naturally reduces levels of carbon dioxide in the atmosphere. In addition, wood has low embodied energy – it does not require much in the way of processing – and it lends itself to recycling.

However, the environmental problems associated with the use of wood are well known. While trees are renewable, ancient forests are not. Deforestation and over-harvesting have endangered certain species and damaged natural habitats, in some cases irrecoverably. In recent years, much attention has been focused on the threat to tropical rainforests, especially in the Amazon Basin and Indonesia, where rapacious, often illegal logging and land clearance have endangered many hardwood species including mahogany, teak, iroko and keruing.

But there has been an equal threat to the world's ancient forests in cold or temperate regions, areas of mixed indigenous woodland that have taken generations to mature and that support a diverse ecology of plants and wildlife. It is estimated that less than 20 per cent of these 'old-growth' forests now remain; that figure is a mere 10 per cent in the case of the United States where old-growth species such as redwood and western red cedar are becoming rare. Thanks to reforestation schemes in many parts of Europe, for example, there are now more trees than there were a century ago, but there is still a need for responsible forestry management. Single-species plantations of softwood, which discourage biodiversity and are more prone to disease, are no substitute for native woodland.

And there are other causes for concern too. Although wood has low embodied energy, timber is often transported vast distances, which adds to its environmental impact. Another potential risk factor has to do with the nature of the material itself. Wood needs protection from fire and pests, but particularly from moisture penetration, and conventional finishing treatments to provide adequate resistance have generally been chemically based – although proper detailing (see page 131) can obviate the need for such treatment. Manufactured wood products, ranging from plywood and particleboard to MDF, are a very efficient way of using timber, because they use waste wood and shavings. However, these products commonly contain urea-formaldehyde binders, which are a proven health risk.

Environmental campaigns have brought many of these issues into prominence. The Forestry Stewardship Council (FSC) is an international body that not only monitors forestry projects worldwide but also seeks to balance ecological considerations with the needs of local communities. Many major furniture manufacturers and housebuilders, including the international furniture chain IKEA, have signed up to schemes whereby they guarantee to acquire timber only from approved sources. FSC-certified foresters and manufacturers meet a set of criteria designed to promote biological diversity and protect native woodlands. Some producers go much further; a number of North American foresters have returned to the old practice of harvesting wood using horses to protect the forest floor from damage by heavy equipment and to minimize damage to the remaining trees.

Types of wood

Wood is an incredibly varied material, in colour, pattern of grain, durability, rate of growth and strength. The basic division is between rapid-growing softwoods, such as pine and fir, and slow-growing hardwoods, which include a diverse range of species ranging from the familiar oak, maple, beech and ash to exotic afromosia, mahogany, teak and sapele.

Softwoods are widely used in construction, both as framing elements and as panels, interior detailing and flooring. Softwoods, however, are less resistant to rot and pests than hardwoods and are generally unsuitable for exterior cladding, with the notable exception of larch which can be used untreated.

Hardwoods are dense, often very attractive and are widely used in furniture, for flooring and as veneers. Many tropical hardwoods are now endangered; most exported Brazilian mahogany, for example, is the result of illegal logging. Particular care should be taken when choosing a

RIGHT Wood can be countrified and rustic or sleek and sophisticated. These beautifully detailed slatted wooden partitions, designed by Glen Murcutt, provide a light screen.

Strategies for the environmentally friendly use of wood:

• Make sure that the timber and wood products you buy come from an approved sustainably managed plantation. Look for the symbol of the Forestry Stewardship Council or similar affiliated bodies.

• Avoid old-growth timber or wood from endangered species.

• Buy timber from a local source, wherever possible. At least seek to minimize the distance the timber is transported.

• Use salvaged or reclaimed wood.

• Choose formaldehyde-free plywood and MDF; choose timber treatments that are as harmless as possible or detail so that timber treatment is not necessary.

OPPOSITE Exposed tree trunk columns (Douglas fir from a managed forest) contrast with the recycled oak floorboards that extend to an outdoor deck in John Broome's self-build low-energy house in southeast London.

LEFT Sophisticated timber panels clad a kitchen area in an eco lodge in Tasmania.

hardwood or hardwood product to ensure that it has come from an approved sustainable source. Threatened tropical hardwoods include ebony, iroko, keruing, mahogany, merbau, wenge, sapele and teak.

Salvaged wood

Using reclaimed or salvaged wood makes good ecological sense as it eases the pressure on existing wood resources. Many reclaimed timber elements, such as doors and floorboards, are thicker, more durable and of better quality than the equivalent new products. Salvaged wood is also available in the form of re-milled timbers or planks, reclaimed from a variety of sources, including old barns, demolished buildings, railway sleepers, and fallen wood.

Salvaging wood can also begin at home. Wherever possible, instead of stripping out old kitchens, cupboards and other types of fitted storage, update with new countertops, drawers and door fronts, leaving the basic carcasses and sub-frames in place.

Manufactured wood products

Between half and two-thirds of all solid timber goes to waste during the sawing and dimensioning processes. Manufactured wood products, however, make use of timber waste and so they represent an efficient use of resources. In addition, most of these products are made from smaller and faster-growing species of trees, which helps to ease the pressure on old-growth forests.

Whereas solid timber varies in strength according to its grain and the presence of defects such as knots, manufactured wood is structurally uniform. In the past, most manufactured woods were weaker than solid timber, which ruled out their use as load-bearing elements. But now there is a new generation of composite wood products, including glulam beams (or bonded spruce planks) and beams made of laminated veneer and laminated boards, which often out-perform solid timber. Such products use adhesives containing formaldehyde or polyurethane but in proportionately far smaller amounts than other types of manufactured wood.

Other manufactured woods include particleboard, MDF, plywood and oriented strand board. Particleboard and MDF both contain a high proportion of formaldehyde, which can off-gas, causing a build-up of pollutants indoors, but both come in formaldehyde-free versions. However, while formaldehyde-free MDF is bonded with lignin, a harmless by-product of cellulose, the bonding agents used to make formaldehyde-free particleboard have their own associated ecological problems. Plywood and oriented strand board (a cheap, strong material made of low-grade timber) both contain formaldehyde but at lower concentrations, while exterior-grade plywood has lower emissions.

Wood laminate is a common and cheap form of 'wood' flooring. In it, an attractive face veneer (or sometimes merely paper printed with a wood pattern) is applied to an engineered softwood core. These products are easy to maintain and durable but often contain formaldehyde binders and high levels of epoxy resins. In addition, many are faced with PVC which is not at all eco-friendly (see pages 150–151). More acceptable are parquet tiles or other types of manufactured-wood flooring systems that consist of a wood veneer over a softwood base. The thickness of the veneer is an important consideration; the thicker the veneer the more often the floor can be re-sanded, so prolonging its useful life and hence conserving valuable resources. The means of fixing is another key factor. Adhesives may off-gas so secret-nailing or snap-and-fit systems are preferable.

ABOVE Timber is a renewable resource. This wooden cabin in the United States, with its wraparound porch and exposed timber structure, blends happily with the wooded setting.

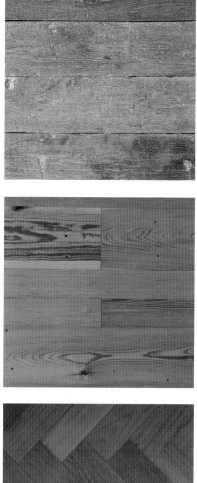

Wood treatment

Wood is naturally prone to being penetrated by moisture and attacked by insects, so defending it against these will mean it needs to be replaced less frequently. The result will be that there will be less pressure on timber resources, which of course is environmentally desirable. Some form of timber treatment is generally necessary for softwoods, whether used externally or in construction, although hardwoods and larch are usually strong enough to not require treatment. However, many timber treatments are chemically based and some are proven toxins, so they pose a problem to the environmentally conscious. The two worst offenders, PCP and Lindane, have been phased out in many parts of the world; others include harmful substances such as arsenic and chromium which can leach out and cause damage. Instead, you should try to use wood preservatives that are derived from borax, since these are the least harmful.

However, design and detailing can help to reduce water penetration: for example, features such as overhanging eaves and drip edges encourage the rapid run-off of rainwater, while wire mesh applied over a timber façade can help to discourage termites. For interior wooden surfaces there are many non-toxic alternatives to chemical preservatives, ranging from beeswax and linseed oil to natural stains and varnishes (see pages 156–159).

OPPOSITE Choose wood responsibly. There are many sustainable options, including oak (top) and pine (centre top). Many exotic hardwoods are endangered. Opt for reclaimed boards or parquet, such as these examples in teak (centre bottom) and muhuhu (bottom).

BELOW White painted wooden tongue-and-groove boarding makes a serene backdrop in a converted villa in Chile.

ABOVE Large plywood panels used as wall cladding provide a warm and tactile surface. Plywood is an efficient use of wood; although it contains formadehyde, concentrations are not as great as in other types of manufactured wood.

RIGHT Original timber columns, new wood flooring and orientated strand board panels provide depth of character in a loft.

ABOVE Woven paper flooring, which orginated in
Japan, is entirely natural and renewable. Paper is
sourced from managed forests, then twisted into
cords and treated with a wax emulsion.

Straw

An agricultural waste product which is traditionally disposed of by burning, straw is finding a surprising new role in both construction and interior detailing. The most familiar form is the straw bale, which is gaining acceptance as a walling material (see page 24). But straw can also be manufactured into non-structural panels and boards.

Straw panels are made by compressing the raw material under high temperature between sheets of heavy paper. The panels can be used to make non load-bearing partitions and ceilings and offer a high degree of soundproofing. Some manufacturers produce panelling systems in classic designs with the panels made of compressed straw and the stiles and rails from formaldehyde-free MDF.

Another useful straw product for environmentally friendly building purposes is straw particleboard, made by mixing chopped straw with a formaldehyde-free binder and pressing it into panels. Straw particleboard is lighter than conventional particleboard and can be used for many of the same applications.

Paper

Paper production accounts for a high proportion of timber consumption so recycling obviously eases demand and conserves resources. Recycled paper is increasingly used for many common paper products, from books and newspapers to wallpaper and toilet paper, but newspaper and paper wasted before and after consumer use can also be processed into materials that perform many of the functions of wood. One American manufacturer produces solid panels, countertops and mouldings made from slurried waste paper without the use of any binders or adhesives. Other types of paper panel are available faced in cork, burlap or fabric for use as wall cladding.

Most wallpapers on the market are now made from recycled paper; in addition, papers are also produced from a range of renewable materials that includes mulberry, hemp and various types of grass. Wallpaper makes an absorbent wall surface and helps to maintain even levels of humidity; plastic- or vinyl-coated papers, however, should be avoided. Because conventional wallpaper paste, although harmless, is a by-product of the chlorine industry, many eco-minded decorators prefer to use starch-based pastes even though adhesion takes longer.

ABOVE Environmentally sound wallpaper is made from unbleached paper, surface-printed with waterbased ink. It contains no PVC or other harmful substances.

LEFT Woven paper flooring is available in a range of subtle, simple weaves for a minimal textured look.

Bamboo

Fast-growing bamboo makes an excellent alternative to wood for many indoor applications, including flooring, panelling, stairs and countertops. Not a true tree, but a woody grass, bamboo grows so rapidly that crops can be harvested every four to six years. There are over a thousand species of bamboo; the types commonly harvested come from managed plantations in the Far East, particularly China. While the use of bamboo entails transportation of the raw material over long distances, its other ecological advantages outweigh the associated energy costs.

Bamboo boards are made of layers of bamboo strips arranged either vertically or horizontally and laminated under high pressure. Vertically laminated boards have a nodular pattern; horizontally laminated boards have a narrower striping. Most, but not all, types of bamboo product contain formaldehyde binders, but some manufacturers are currently investigating more environmentally friendly adhesives.

Bamboo is stronger than oak, maple and beech, and very stable, which makes it less prone to expansion and contraction caused by variations in temperature and humidity. It comes in a range of shades from natural blonde through to deeper and warmer amber colours and can be finished with the same treatments as wood.

Cork

Derived from the outer bark of the evergreen cork oak (*Quercus suber*), a tree native to Spain, Portugal, southern France, north Africa and Italy, cork is a natural, healthy material that makes a warm, resilient surface for floors or walls. Every decade the cork oak naturally sheds its thick outer bark, which means that the cork can be harvested without any damage to the tree. This is a process that also entails very little wastage. Binders are required to hold the individual granules of cork together and, while adhesives containing formaldehyde were once common, less harmful binders are often used today.

Cork has many advantages: it naturally resists rot and mould and it also makes a great sound-absorber and insulator. Produced in sheets, rolls, slabs or tiles for a variety of applications, cork typically comes in warm honey shades, although some stronger colours are available. Cork tiles or sheet used for flooring need protection; wax is an eco alternative to conventional polyurethane seals.

ABOVE Bamboo is gaining popularity as an eco alternative to hardwood. It is suitable for a wide range of applications, from flooring to countertops.

RIGHT Derived from the bark of the cork oak, and harvested in a process which causes no damage to the tree, cork is a cheap, resilient and healthy material.

Linoleum

Linoleum is a wholly natural product with very good environmental credentials. Its raw ingredients are widely available and renewable and the processing entails no release of toxic gases. It is made from linseed oil, a by-product of flax (*oleum lini*, hence its name), pine resin, powdered cork, wood flour, powdered limestone and pigment, pressed onto a jute or hessian backing and baked at high temperature. Because it is anti-bacterial it is ideal in kitchens, bathrooms, family rooms or wherever an easy-care hygienic floor is required. And being anti-static too, it repels the dust that attracts mites and so is good in homes where asthmatics or those suffering from allergic reactions live. It can also be used as a surface for desks or tabletops.

As a floor, it is warm, matt, resilient and comfortable underfoot. Today, it is available in a range of strong colours, typically with a matt, mottled appearance, and can be used to create graphic floor-level designs using sophisticated computer-controlled cutting techniques. Linoleum is installed using water-based environmentally compatible adhesive.

Linoleum also scores well in terms of performance and actually gets tougher as it matures with age. This maturing process, which involves the oxidation of lineolic acid, does entail the off-gassing of VOCs (see page 12). However, these are believed to be much less harmful than those off-gassed from vinyl. Linoleum is far preferable to vinyl from an environmentally friendly point of view and can be used for the same applications.

Rubber

Our continuing love affair with the car results in mountains of discarded car tyres dumped on landfill sites around the world. In the United States alone, nearly 250 million tyres are thrown away every year. One answer to this blight is the recycling of tyre rubber to produce rubber flooring.

This is as strong and resilient as the original tyre rubber and is highly slip- and weather-resistant. Inexpensive, tough and available in a range of strong colours and surface patterns, it is ideal for areas of heavy traffic, particularly indoor/outdoor areas and entranceways. Concerns, however, have been raised about its effect on indoor-air quality. While manufacturers stress that recycled rubber surfaces have low VOC emissions and meet indoor-air quality guidelines, some experts advise against their use in the home.

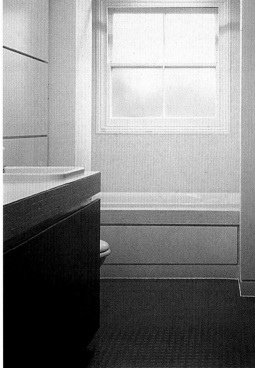

LEFT AND OPPOSITE Linoleum is a natural, hard-wearing, anti-bacterial and hypo-allergenic product. It comes in in a wide range of colours, patterns and textures and is an excellent alternative to vinyl.

ABOVE Flooring is now available made of recycled tyre rubber, although there are concerns about domestic use due to potential VOC emissions.

Stone

As a natural material, stone has many admirable qualities, both aesthetically and practically speaking. Although stone is generally expensive, most types are incredibly durable; stone construction, roofing, flooring, worktops and other interior surfaces have the potential to last for many generations and, in the right context, with a minimum of maintenance. In terms of appearance, natural stone varies widely in colour, texture and surface patterning, from dark sleek slate to cool sophisticated limestone through to exotically veined marble. Stone also lends a sense of permanance and undeniable luxury to the interior.

In eco terms, one of the great advantages of stone – along with brick, concrete and other types of masonry – is that it stores heat well and can play an important role in houses designed for passive solar gain. In areas of southern Europe, for example, thick masonry walls and stone or terracotta floors help maintain moderate indoor temperatures year round, the dense solid materials gaining heat slowly during the summer months and releasing that heat gradually over the winter. In northern regions, stone floors, walls or other interior surfaces add thermal mass to timber-frame buildings which would otherwise heat up and cool too quickly.

Unlike wood, however, stone is a natural resource which is neither renewable nor inexhaustible. Types of stone that have historically been highly sought after for their practical qualities, beauty, patterning or coloration, such as certain varieties of marble, may be rare or no longer available. Yorkstone, a type of sandstone found in Yorkshire, which has long been in demand for outdoor paving because of its weather-resistance, is on the verge of becoming rare.

While compared to metals and plastics relatively little energy is consumed in the extraction and processing of stone, energy costs can be high where stone is imported from regions halfway across the world. Deposits of different types of stone are not uniformly distributed, with the result that most stone travels a significant distance from origin to point of use. In the case of rare or unusual varieties, that distance may be thousands of miles. The associated energy costs are enhanced by the weight of the material.

Another and perhaps more obvious environmental problem is the impact that quarries and mines have on natural habitats. In the past, over-exploitation of certain types of stone has ravaged landscapes all over the world, leaving them permanently scarred.

Types of stone

Stone falls into three basic categories, according to how and when it was formed. Hard, dense igneous rocks, such as granite, are the oldest, created when molten rock cooled and crystallized during the formation of the earth's crust thousands of millions of years ago. Sedimentary rocks, such as sandstone and limestone, date from more recent geological periods and are the result of deposits laid down by rivers, lakes and seas. These types of stone are less wear-resistant and softer than igneous rock. The last and youngest category is metamorphic rock, rock that was subject to intense heat and pressure when mountain ranges were formed during the movement of the earth's crust. Marble and slate are metamorphic rocks.

Stone has a wide range of applications in the interior, with the most common being as flooring and as kitchen worktops. Stone flooring makes particular sense in hallways and entrances that see heavy traffic, in indoor/outdoor areas such as garden rooms and conservatories and in rooms such as kitchens and bathrooms where water-resistance is important. Stone bedded in sand and cement mortar is more environmentally sound than stone installed with adhesive. For all flooring applications, the sub-floor must be sufficiently strong to support the weight of the material.

In the case of worktops, the type of stone you use is critical. Very porous stone, such as limestone is not suitable because it stains readily, particularly when it comes into contact with acids such as wine, citrus juice and vinegar. Hard dense stone, such as granite, makes a more practical and wear-resistant worksurface.

Local stone

Much of the inherent 'rightness' of vernacular buildings derives from the fact that generally speaking they are constructed out of local materials. In the past, only the extremely wealthy could afford to have exotic imported stone in their homes; ordinary people used what was to hand – either stone quarried in the immediate vicinity or stone that had simply been cleared from the land. Today, local stone is the best source for those seeking to minimize energy costs; in areas where stone is not readily available, 'local' may be defined as 'regional' or 'national'.

Salvaged stone

Reclaimed or salvaged stone is an excellent choice for people looking for an environmentally friendly alternative to new quarried stone. Because stone is so amazingly durable, re-use

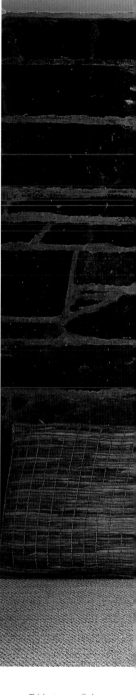

ABOVE Thick stone walls have great thermal mass. Stone quarried locally is a feature of vernacular buildings; the exposed surface has great depth of character.

makes good sense from the practical point of view; most types of stone also acquire a pleasing patina with time, and the resulting depth of character is much sought after. In fact, antique stone salvaged from old farmhouses, monasteries, churches and manors often commands a premium price. The places to look for salvaged stone include architectural salvage yards, demolition sites and monumental masons (for offcuts of marble and granite). Roofing slates can be re-used as floor tiles, provided they are laid on a concrete screed. As with new stone, avoid salvaged stone that has been imported or transported long distances.

Strategies for the environmentally friendly use of stone:

• Avoid imported stone. Seek out local sources and attempt to mimimize transportation distances wherever possible.

• Choose reclaimed or salvaged stone.

• Use the thermal mass of stone to enhance passive solar strategies, for example by installing stone flooring in glazed areas or near large windows.

• In cooler climates, underfloor heating, which is an efficient form of space heating, can make stone floors more comfortable.

Brick

Brick is one of the oldest, if not the oldest, manmade building material known. Its raw ingredient, earth, or more specifically clay, is found all over the world. The earliest bricks were little more than clay mixed with straw or some other form of natural binder and baked hard in the sun – a basic technology that survives in the form of the adobe brick.

Like stone, any earth-based materials have high thermal mass. When they are warmed by the sun they store the heat and then release it slowly back into the interior. Unlike stone, however, manufactured brick has relatively high embodied energy – because of the extreme heat required for the firing process – and this factor must be offset against any energy savings that can be made when it is used. Salvaged brick, which is widely available although occasionally more expensive than new, is an eco option.

Brick is a standard construction material in some areas of the world, particularly in central Europe. Nowadays new types of brick have been developed which are honeycombed with air pockets. This is a feature that both lightens the weight of the brick and dramatically improves its thermal insulation qualities.

As an internal building element, brick has many positive advantages to recommend it. Floors and walls made of brick are both warming and warm-looking. They are easy, unpretentious surfaces that have an inherent domestic appeal. In addition, thanks to their porosity, exposed or unplastered brick walls naturally help to regulate levels of humidity in a building. Brick floors require little in the way of maintenance, although because of their porous nature you should take extra care to avoid spills; despite their absorbency, sealing is not a recommended finish. They combine well with underfloor heating systems.

ABOVE Built of recycled brick and wood, this house in Bogotá, Colombia, is integrated into the landscape to provide shelter from the cold Andean winds. The long sides run east/west to maximize light; thick brick walls have high thermal mass.

ABOVE RIGHT Handmade terracotta tiles make a stylish and practical waterproof surface for all wet areas. To prevent excess condensation, leave a portion of the wall area untiled.

Cheap, durable and extremely wear-resistant, quarry tiles were first manufactured in the nineteenth century as an alternative to the more traditional terracotta tile. As with terracotta tiles, the basic ingredient is clay, in this case unrefined high-silica clay, which is pressed into a mould and burnt. Less characterful than terracotta, quarry tiles nevertheless score well on both practical and environmental grounds. Like terracotta tile and brick, quarry tiles work well with underfloor heating.

Ceramic tiles

Highly processed ceramic tiles come in a wide range of colours, sizes, textures and degrees of durability. The basic ingredient is refined ground clay, which is pressed and fired at high temperature. Regular and contemporary in appearance, ceramic tiles make a no-nonsense surface for floors, walls and splashbacks, depending on the type. Fully vitrified ceramic tiles can be used outdoors. When used for flooring, ceramic tiles can be combined with underfloor heating to increase comfort.

Although fully tiled wet areas, such as bathrooms, look better visually, they can raise levels of condensation and humidity unacceptably. Eco designers advise restricting tiling only to those areas immediately in need of water-protection, so that walls have the opportunity to breathe and absorb moisture. Many tile adhesives have a high VOC content, but effective water-based alternatives are available.

Terracotta and quarry tiles

Terracotta tiles are appealing natural earthen products. While most terracotta tiles are now mass-produced, handmade tiles, produced in areas of the world such as Tuscany, Provence and Mexico, are also available, and are much more characterful. This is thanks to the variations that exist in the composition of the local clay deposits, along with irregularities in the shaping and firing – generally done in a wood-fired kiln. Antique terracotta tiles, beautifully distressed by time, are another option, although prices tend to be high. Both handmade and manufactured terracotta tiles have a leathery patina which deepens over time. Used chiefly as a flooring material, unglazed terracotta requires sealing with linseed oil or wax for full protection. Glazed terracotta, although not as robust as ceramic tiles, comes in a wide range of colours, patterns and motifs.

Concrete

Superficially, concrete might appear to be the antithesis of an eco material. The sheer brutality of concrete and its widespread urban use in the postwar period to create unlovely utilitarian structures has given it an image problem only minimally eroded by its recent cachet as the favoured material for cutting-edge minimalists.

Because of its strength and resistance to moisture, concrete is unavoidable in certain applications, notably in foundations. But due to its high thermal mass, concrete can also play a more positive part in eco design. Like brick and stone, it helps conserve energy, particularly where it is used as a flooring or sub-flooring or to create internal walls. In addition, concrete is naturally moisture- and insect-resistant, which means that it requires little in the way of maintenance,

Strategies for the environmentally friendly use of concrete:

- Build lightweight structures that require minimal concrete foundations.
- Replace fly ash for a proportion of the Portland cement in the concrete mix.
- Use self-coloured concrete as a final finish.
- Use lightweight concrete blocks.
- Exploit concrete's high thermal mass for passive heating and cooling.

and by adding pigments to the basic mix it can be used as a final finish without the need for additional toppings or special paints, which may not be environmentally friendly. Another plus from the environmental point of view is that the ingredients that go to make up concrete – sand, gravel and Portland cement – are abundant and widely available.

On the down side, the manufacture of one of concrete's key ingredients, Portland cement, releases significant amounts of carbon dioxide into the atmosphere, and thus contributes to global warming. Additionally, gravel extraction can damage natural habitats while concrete formwork – the framing typically made of wood that is used to provide a mould for concrete structures – often ends up discarded, so wasting natural resources.

There are various ways of improving concrete's eco credentials, both in terms of its composition and the way it is employed. One is to use fly ash in place of Portland cement. Fly ash is slag or waste from coal-fired power plants. It can be used as a substitute for between 25 and 60 per cent of the Portland cement, depending on the strength required. Another effective way of minimizing the eco impact of concrete is to use less of it; for example, by specifying honeycombed or lightweight concrete blocks. These also have a higher insulating rating, which is an eco bonus.

Terrazzo

A luxurious and sophisticated material, terrazzo is an aggregate of marble, granite or stone chippings, mixed with concrete or cement and poured or laid as tiles. A new eco alternative is a terrazzo lookalike made from recycled glass and non-toxic resins. This can be polished to a high shine and used indoors and out.

Plaster

Plaster is one of the most common internal finishes for walls and ceilings; external plastering, or stucco, is also widely used as render for masonry or brick walls. The basic ingredients of plaster are sand plus a binder, which is usually either lime, cement or gypsum. Lime and cement plasters are generally used externally, while gypsum, a mineral in a range of colours from white and grey to red and yellowish-brown, is a common component of interior plaster and plasterboard or drywall. Gypsum plaster is easier to work than lime plaster.

From the environmentally friendly point of view, lime- and gypsum-based plasters are a better choice than cement-based varieties because they are more permeable to both heat and moisture. This means that they enhance the ability of the structure to 'breathe' and so help prevent the build-up of condensation.

Plastered internal surfaces can be left bare, sealed with no more than a coat of wax to prevent surface dusting, or painted with a natural or eco-friendly paint. Pigments or other additives can be applied to the basic plaster mix to give soft, diffused colour or textural variation.

Traditional plasters, made with mud, or coarse sand, animal hair and lime, are increasingly used as external renders for straw bale or earthen houses. An eco alternative to conventional plasterboard is wallboard made of recycled newspaper and gypsum over a recycled gypsum core.

OPPOSITE During the daytime, a concrete floor will absorb heat from the sunlight pouring through the windows, and then release the stored heat slowly overnight.

BELOW The beautifully textured clay plastered walls in this adobe house need no subsequent finishing. Clay, mud and lime plasters make breathable surfaces.

Glass

The basic ingredients of glass – sand, soda and lime – are abundant and widely available but the manufacturing process itself, which entails melting the ingredients at temperatures of more than 1,500 degrees centigrade, consumes huge amounts of energy. However, glass can be recycled very easily and without any decrease in quality.

In almost all homes, except perhaps those in the hottest or driest regions, glass is indispensable, protecting from the elements while admitting natural light into the interior – essential for physical and psychological well-being and for reducing dependence on artificial light. Until relatively recently though, expanses of glass dramatically compromised a building's energy efficiency, overheating the interior during the warm sunny months and draining heat during the winter or at night. Nowadays, however, considerable advances have been made in glass technology, enabling quite significant areas of glazing to be incorporated into walls and roofs without the associated impact on energy consumption.

Low-emissivity glass

Today's developments in glass manufacture reflect the need for better environmental performance. Ordinary glass is brittle, shatters readily and transmits heat, but recent advances in glass technology have led to the development of low-emissivity (low-E) glass. This is coated with many thin transparent layers of silver oxide that reflect infrared energy back into the interior, so dramatically reducing heat loss. Low-E glass has a U-value that approximates to that of a well-insulated cavity wall; multiple-layer high-performance units can be up to three times more insulating. For hot climates, there is a type of low-E glass where the coatings are used to reflect radiant energy before it reaches the interior.

Glazed units

Double- and triple-glazed units, where panes are separated by an air- or gas-filled cavity or cavities, are highly insulating. Where low-E glass is used, cavities are typically filled with argon or krypton. Some extremely high-performance units incorporate venetian blinds between the panes which can be adjusted by controls on the outside.

Window frames

A considerable proportion of any window opening is actually the frame rather than the glazing, so both the material that is used to make the frame and the way it is detailed and designed is bound to have an impact on performance. The three materials most commonly used to make window frames are wood, aluminium and PVC. Wood has the best insulating properties of all, PVC is the cheapest and aluminium is the most long-lasting.

Most of the wood used in window frames is softwood (often old-growth pine) which needs to be regularly repainted to provide weather-resistance. At the expensive end of the market tropical hardwoods, such as teak and mahogany, are also employed, because of their durability and the fact that they weather well naturally without the need for protective coatings. Although new windows with hardwood frames should be avoided on ecological grounds, any existing hardwood frames should be re-used and if possible upgraded with low-E glass or double glazing.

Aluminium frames are very durable and require little maintenance, but because of aluminium's high embodied energy, they are not recommended. Aluminium frames also dramatically decrease the window's insulating properties.

Because of its cheapness, most commercially available windows are manufactured from extruded PVC, which has a good insulating value and can be fashioned into a huge range of designs and styles. However, the use of PVC is strongly discouraged on environmental and health grounds. PVC is the most environmentally problematic plastic and one of the world's largest sources of harmful dioxins.

Given all these pros and cons, the most environmentally friendly choice of material for window frames would be locally sourced wood, but this can be difficult to obtain and verify. A good compromise, marrying the sustainability and insulating properties of wood, with the low maintenance of metal, is a composite frame made of timber section clad with powder-coated aluminium.

Many double- or triple-glazed units have hollow frames, with added insulation so the frame has a similar insulating value to the glass. The joint between panes and frame is sealed with silicone for complete weather-tightness.

RIGHT Technological developments in glass production mean that interiors can be flooded with natural light without excessive heat gain or loss. This Californian ranch house is glazed with low-E glass.

Strategies for the environmentally friendly use of glass:

- Position windows to benefit from solar gain and natural light.
- Specify low-emissivity glass to reduce heat loss, particularly for areas of extensive glazing, such as glass roofs.
- Double- or triple-glazed windows.
- Avoid PVC and all-metal window frames.

Metal

Sleek and sophisticated, with an industrial, utilitarian edge, metal has found a widespread use in modern building, both as a structural element and interior finish. Steel, virtually synonomous with the towering form of the skyscraper, has increasingly been favoured as an up-to-the-minute cladding for kitchen counters and units. Aside from aesthetic considerations, steel and other metals such as copper, iron and zinc, are also strong, durable and pest-resistant.

In eco terms, however, the use of metal can be very problematic. Of all the materials commonly employed in building, metal has the highest embodied energy, some 300 times that of timber. Although metals are generally plentiful, with the exception of zinc and tin which are becoming scarce,

their extraction and mining can despoil habitats, while metal processing is a significant polluter. Additionally, because metals conduct heat so well, metal framing, for example in windows, can create cold bridges (see pages 28–31) that compromise a building's energy efficiency and encourage condensation and mould.

Set against these significant drawbacks is the ease with which metals can be recycled and, because they are relatively expensive, their recycling is better established than that of many other materials. Thus one-third of all aluminium is recycled; half the iron needed for steelmaking is scrap and the steel industry overall has a 68 per cent recycling rate. Steel with a recycled content of up to 90 per cent is available in products as diverse as structural beams and nails and some designers are re-using steel beams in new buildings.

ABOVE Greenwich Millennium Village in London is a development of 'green' homes on a brownfield site. The houses are made from lightweight prefabricated materials, with roofs of corrugated aluminium to help collect rainwater.

RIGHT Steel is so strong that supporting elements can be relatively minimal, as in this Australian design by Glen Murcutt. The metal reflects the heat of the sun.

Strategies for the environmentally friendly use of metal:

• Use structurally in light sections, and bolted rather than welded together to facilitate re-use.
• Use as fasteners, handles and other forms of ironmongery in place of synthetic materials.
• Use recycled metal and recycled metal elements wherever possible.

An advantage of recycled metal is that it has a dramatically lower embodied energy: recycled aluminium, for example, consumes only 5 per cent of the energy used to extract and process aluminium from scratch.

Although some eco designers eschew the use of metals for all but the most minimal applications, others believe that metals can play an important role when it comes to enhancing the durability and stability of a structure or surface, particularly in the form of connections, fasteners and weatherproof cladding. Rolled corrugated iron sheet, for instance, a common Australian industrial product, is employed by architects such as the Australian Glen Murcutt as roofing, the corrugations promoting rainwater collection and the reflectiveness of the material helping to prevent interiors from overheating. Although their embodied energy is comparatively high, metals can also be used in light, efficient sections so that less is needed to do the same structural work. Another eco-advantage is that if bolted together rather than welded, metal framing can be de-mounted and re-used at a later date.

More direct recycling, for example by re-using metal that has been salvaged from retail or commercial applications, can be a useful strategy for acquiring metal surfaces and finishes for the interior. Scrapyards, junk shops and architectural salvage companies can be a good source for metal-clad or all-metal fittings and fixtures, such as old office desks, filing cabinets, secondhand catering equipment, industrial racking systems, lockers and the like, all of which can provide robust no-nonsense storage for the home. Taking the salvage strategy to the extreme, LOT/EK, a New York-based architectural practice, uses heavy-duty scrap – petrol tanks, cement mixers and shipping containers – to create individual living modules.

Strategies for the environmentally friendly use of plastic:

- Avoid new plastic products and materials wherever you can.
- Avoid the use of glues, seals and other finishes that contain plastics such as epoxy resin and formaldehyde.
- Refuse plastic packaging.
- Use recycled plastic goods.

LEFT Translucent colourful recycled plastic has been used to create these striking floorstanding lights.

Plastic

Plastic is the ultimate modern material. Light, cheap, colourful and convenient, it can be engineered to suit almost any performance requirement. Embracing a huge family of related materials that differ widely in composition and characteristics, plastic has infiltrated our lives to an extent unimaginable half a century ago. Since the 1950s, worldwide production of plastics has soared from less than 5 million tonnes annually to over 80 million tonnes.

But the runaway success of plastic has been bought at the expense of a massive increase in waste, most of which ends up buried on landfill sites. Britain annually consumes 3.5 million tonnes of plastic, discarding 2.5 million tonnes as waste. Over 60 per cent of that waste is packaging: 8 billion plastic carrier bags are thrown away in Britain every year. In the United States, discarded plastic toothbrushes alone account for 27 million kilograms of plastic waste a year.

As well as the waste, there are huge environmental problems associated with the production of plastic and increasing concern about its effect on human health. In ecological terms, plastic is simply the most expensive material there is.

Although the raw ingredients that go to make up most types of plastic – petroleum and natural gas – are technically organic, plastic is so highly processed that it cannot be considered a natural material in any sense; moreover, many types are not biodegradable. In addition the production of plastics consumes vast amounts of energy and is significantly polluting, and most types of plastic are highly flammable and give off toxic smoke when they burn.

Although there are over 50 different types of plastic, one of the most common in the building industry is PVC or polyvinyl chloride which accounts for 16 per cent of all synthetics produced annually. PVC is found in many behind-the-scenes applications, including pipes, cisterns and as a coating for cabling and wires, as well as more obvious finishes such as vinyl flooring and window frames.

In addition to the ecological threat posed by the increased reliance on PVC, there are significant concerns about its impact on human health and the risks associated with it have led to it being banned in some parts of the world. PVC readily off-gases into the air and has been linked to immune system and nervous disorders; when it burns, it gives off highly toxic, dioxin-laden smoke. However, even environmentalists acknowledge that in the case of some applications – for example, rainwater and waste gutters and pipes – PVC cannot be replaced economically.

Although it is not always possible to avoid all use of plastic in the home, it is certainly worth making do with as little as possible. In many cases, particularly for surfaces and finishes, alternative materials are readily available: use linoleum instead of vinyl flooring, for example, or wood instead of laminate countertops.

Types of plastic

Many different types of plastic are commonly found in the home. In some cases, such as plastic containers, plastic shower curtains and vinyl flooring, the plastic content is obvious and easy to identify. In other cases, it may not be so immediately apparent. Various types of glues, seals and resins, such as formaldehyde, epoxy resin and polyurethane which are widely found in solid materials such as laminate and composite boards and in insulating materials and foam, are types of plastic. Many of these are known health hazards.

Acrylic is a rigid sheet material widely used as a substitute for glazing or to make shower cubicles, baths and sinks. It is highly flammable.

Polythene is used in packaging, to coat metals (particularly aluminium) and to insulate cables.

PVC has a huge range of applications, from vinyl flooring to damp-proof membranes, shower curtains and inflatable furniture.

Polypropylene is used in many household objects, such as bins, brushes, and chairs.

Polystyrene is common in packaging, but is also used as a substitute for plaster for ready-made decorative covings and details.

Nylon is found in many types of cheap carpet; also used to make curtain rails and door furniture.

Polyvinyl acetate is found in emulsion paint, floor finishes and adhesives.

Melamine is used to make laminates for worksurfaces and counters.

Polyurethane is widely used both in paints and varnishes and to make insulating materials and foam mattress or cushion linings.

Epoxy resins are used to make glues and coatings.

Formaldehyde is used as a preservative or binder in fabrics, carpet and manufactured wood products.

Ureaformaldehyde is found in glues and floor seals.

Recycled plastic

One of the significant difficulties associated with plastic is that its cheapness has encouraged us to use it as a disposable material rather than recycle and re-use it. There are, however, encouraging signs that the situation is changing. One response to the problem has simply been to use less plastic: today's yoghurt cartons, for example, are half the weight that they were in the mid-1960s. The other response is to recycle plastics. In basic terms, recycling can simply mean re-use – returning a container to a supplier to be refilled, for example, or using a carrier bag until it wears out. Increasingly, plastics are being collected and recycled into new products. A large proportion of carrier bags and clothes hangers are now recycled, either as bags and hangers, or in the form of sheet materials that can be jointed and worked like wood.

Over the last decade the number of businesses specializing in recycling plastics has trebled and technological developments have improved both the recycling process and the quality of the end products. The fact that there are many different types of plastic has posed a problem for recycling, but mandatory labelling has helped make sorting by type easier. Today, a vast range of goods are produced from recycled plastics: bin liners, carrier bags, refuse sacks, bottles, flooring, sheeting, CD cases, fencing – even textiles, such as fleece. One of the most surprising examples is a pencil whose body and lead are produced from recycled vending-cup waste. In a striking example of 'closing the loop', one American company produces toothbrushes out of plastic recycled from the tubs used to transport computer chips; the toothbrushes are supplied with a prepaid envelope so they can be returned to the company to be recycled again when they have worn out.

BELOW Recycled plastic clothes hangers have a tortoiseshell appearance when turned into plastic sheet.

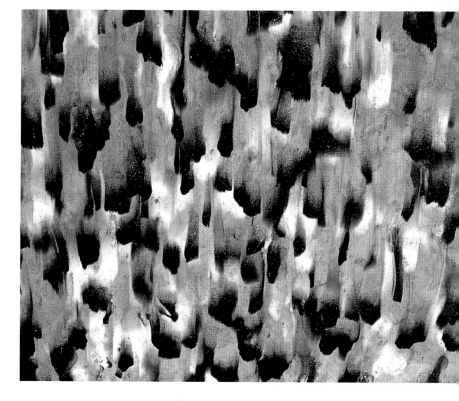

Natural fabrics and weaves

OPPOSITE Wholly natural wool products, from blankets and throws to cushion covers and carpeting, are untreated with pesticides and dyed using organic vegetable dyes.

Soft furnishings, in the form of fabrics and weaves, spell comfort, warmth and tactility. Even the most spartan of minimalists finds some use for fabrics in the home, if only as bed or bath linen; for most people, carpets, rugs, curtains and other types of furnishing are an important way of making a home pleasant and livable.

Natural fibres, from cotton and linen to wool, jute, sisal, coir and hemp come from readily available and renewable sources and are biodegradable and recyclable. Unlike the manufacture of synthetic materials, the processing of natural fibres demands little energy.

While it is relatively straightforward to avoid the use of synthetic materials – a simple label check is usually enough to ascertain the composition of a fabric or weave – many ostensibly natural products have been treated with potentially harmful chemicals, either during the growing cycle or as applied finishes, and these chemicals can cause damage both to the environment and human health. Flame-retardants, for example, are often applied to clothing, mattress or upholstery materials (and are a legal requirement in many parts of the world) but they are a health hazard. Another problem, which arises in the case of carpeting and other types of soft flooring, is that natural materials may be backed with synthetic ones to improve resilience and wear.

Types of natural fibre

When choosing fabrics or woven materials, it is important to check labels carefully to make sure they are synthetic-free. Additives can be hard to spot, although fabrics marketed as 'crease-resistant' or 'easy-care' have often been treated with formaldehyde and should be avoided. It is also important to select the right weight and type of material for the application: lightweight fabric, for example, used for upholstery will only wear out faster and require replacement.

Cotton is a natural fibre that is derived from the cotton plant. Unlike synthetic materials which can feel unpleasant, cotton breathes and absorbs moisture, keeping body temperature even and so making the fabric comfortable next to the skin.

Unfortunately, because the cotton plant is vulnerable to a large number of pests, pesticides, fungicides and fertilizers are widely used by cotton growers and residues may remain in the fabric. Most cotton is subsequently chemically bleached; coloured cotton is chemically dyed and fixed.

Both unbleached and organic cotton made up into bed linens, futons and clothing is increasingly available from specialist suppliers. Unbleached cotton can be bleached naturally in the sun and dyed with vegetable dyes if required. A new type of cotton fibre has also been developed recently that produces green and brown cotton naturally without the need for dyeing.

Linen comes from the flax plant. Like cotton, it is very absorbent and it is also cool and extremely strong. Pesticides are used in flax-growing but not to the same extent as in the growing of cotton. Most linen is now chemically bleached, and chemical dyes and other additives such as flame-retardants are commonly used. Choose natural, pale or unbleached linen. If you wish, you can bleach it in sunlight and dye it with vegetable dyes.

Wool comes from a range of sources, including goats, alpacas and vicuñas, as well as sheep; felt is made from matted wool fibres. Absorbent, warm and more flame-resistant than cotton or linen, wool is manufactured with little use of chemicals, but it is often treated with a pesticide for mothproofing. A significant proportion of wool today comes from Australia, which means that there will be energy costs involved in its transportation to other parts of the world. For an eco-friendly approach, look for untreated wool, unbleached wool or wool dyed with vegetable dyes. Sheepskin fleeces allow air to circulate round the body and so make excellent underblankets, particularly for babies, children or the bedridden.

Silk derives from the fibres of silkworm cocoons. Light, smooth and supremely luxurious, silk takes dye readily and is available in a range of intense hues. However, silk is often treated with mothproofing chemicals and, because most silk is imported, there are the inevitable energy costs involved in the transportation.

Reeds and rattan Grasses and woody plants are natural and renewable sources of fibres that can be woven into matting, baskets and chair seats. The problem with both reeds and rattan (which comes from the southeast-Asian rattan palm) is over-harvesting and habitat destruction, so care should be taken to source products and materials from sustainably managed plantations.

Sisal is a strong fibre woven from a species of agave and typically made into mats and floor coverings. One of the most popular of the natural fibre floor coverings, sisal is hard-wearing enough for areas of heavy traffic but not as rough to

walk on as coir. It can be dyed and is available in a range of patterns, but it is not water- or stain-resistant.

Seagrass comes from a grass native to China. Tough, cheap and anti-static, seagrass weaves are smooth and comfortable underfoot. Naturally water-resistant, it does not stain. It cannot be dyed.

Coir comes from the coconut husk. Coir weaves are made into floor coverings and doormats. It is very hard-wearing, but feels scratchy to the skin.

Jute comes from a subtropical plant native to India and has long been used in rope-making, for making coarse cloth such as hessian, and for backing carpet and linoleum. More recently, it has been employed to make floor coverings. Jute weaves are soft and much less durable than sisal or coir. Because demand for jute exceeds supply, similar fibres such as ramie are often used as a substitute, particularly for making coarse cloth such as hessian.

Hemp is a rather coarse fibre that is derived from the cannabis plant and is used to make matting and cloth as well as rope. Because the growing of cannabis is strictly controlled, for obvious reasons, the full exploitation of this fibre has been held back.

Rayon Although rayon is a synthetic fibre, it is made from cellulose rather then petrochemicals. It is commonly used in blends with natural fibres.

Strategies for the environmentally friendly use of fabrics and weaves:

• Check labels carefully to determine the fabric's content is wholly natural.

• Avoid fabrics or weaves that have been chemically treated to improve wear or maintenance.

• Buy organic unbleached cotton or linen.

• Use natural underlays for carpet and/or other natural fibre floorings.

Natural-fibre flooring is non-oily, which makes it hypo-allergenic. Made from renewable sources, such as sisal, jute and coir, it is laid like carpet but is not fully stain-resistant. Varieties include herringbone-weave seagrass (above), coir (top right), large bouclé sisal (centre right) and small bouclé sisal (below right) dyed in three colours.

Carpeting

Carpet is warm, resilient and sound-proofing, but most is made of synthetic fibres or synthetic/natural blends, while all-wool carpeting is often treated with moth-repellent and other chemicals during manufacture. Most conventional carpets are also backed with synthetic materials, such as latex. Eco alternatives are available, but may be difficult to source. The best option is natural-fibre floor coverings such as sisal, jute or coir. Carpet also attracts dust mites – a major cause of asthma and other allergenic reactions – so avoid wall-to-wall carpeting if there is an asthma sufferer in the household.

Underlay improves durability, but many underlays are made of synthetic materials. Alternatives include cushioning or underlay made of jute and animal hair or of recycled and waste synthetic materials. It is better to lay carpet by stretching and tacking than by gluing.

The enduring popularity of carpeting has led to a corresponding problem with waste. Over two million tonnes of carpet are thrown away every year, with most ending up on landfill sites. New closed-loop carpet leasing schemes are gaining in popularity; in these schemes, carpets are available to rent from manufacturers, who take the carpet back at the end of its useful life and recycle it into new carpet.

Upholstery

Upholstered furniture, sofas and easy chairs, often contain a high proportion of polyurethane foam, which is harmful both to the health and the environment. Alternatives include organic cotton batting or various mixtures of traditional upholstery materials including wool, sisal and coconut fibre, along with feather- or down-filled cushions.

Mattresses and bedding

The bed, or more specifically the mattress, is the most important piece of furniture in the home. We spend almost a third of our lives in bed and it only makes sense to ensure that your mattress is as healthy as possible. Natural mattress and bedding materials absorb moisture and allow the skin to breathe. However, in many countries strict regulations with respect to fire-retardance are in operation; in the United States, for example, it is necessary to produce a doctor's letter certifying sensitivity to chemical in order to purchase an organic cotton mattress which contains no fire-retardant chemicals. Organic cotton pillows and futons are also available. Feathers and down are the most environmentally sound fillings for duvets and these are not treated with any chemicals during the preliminary cleaning.

ABOVE Organic cotton products, such as this cotton-covered duvet, are not treated with flame-retardant agents and are tested for chemical residues.

Paints, varnishes and seals

Finishes, including paints, varnishes and seals, have a practical as well as a decorative role to play. Many types of material last longer if they are properly finished, particularly wood which is susceptible to damage from excess moisture. However, the most commonly available finishing products today contain toxic chemicals and a high proportion of plastic. Painting a wall with ordinary emulsion paint, for example, is akin to coating it with a thin layer of plastic, which essentially prevents the underlying plaster from breathing. The healthy properties of natural surfaces such as wood, stone, brick, tile or plaster can be severely compromised by the application of chemically based finishes and seals and, depending on the surface area, the impact can be considerable.

The production of conventional paints and finishes is particularly wasteful. A typical polyurethane, for example, is produced using high-energy petrochemical catalytic cracking processes, the result of which is approximately 10 per cent product. The remaining 90 per cent ends up as waste and must be stored in tanks because we do not know at the present time how to dispose of it safely. A natural paint, on the other hand, is mostly product and any waste generated by production is biodegradable.

With growing concerns about these issues and about the risks of VOCs, which are found in most commercial paints, seals and varnishes and which adversely affect indoor air quality, increasing numbers of producers now market natural finishes containing vegetable or mineral pigments and natural solvents, while standard paint manufacturers have hopped on the bandwagon and introduced ranges of 'low-odour' or 'low-VOC' paints.

However, all these products have attracted a certain degree of controversy, both on environmental and health grounds. For example, 'low-odour' or 'low-VOC' paints may contain different harmful chemicals, while their absence of odour can remove a warning signal which would otherwise alert the user to potential risks.

In addition, there is a question mark over the effects of the natural resin oils used as solvents in natural paints, with some experts identifying such ingredients as allergens in their own right. In practical terms, natural paints are also more difficult to apply and have longer drying times.

In general, however, it is important to avoid synthetic finishes wherever possible, if only because of the impact on the environment caused by their production.

Types of finish

One advantage of natural paints is that their ingredients are clearly listed on the outside of the container. Standard commercial paints may contain a wide range of undeclared chemicals, including formaldehyde, fungicides and bactericides, as well as heavy metals. Most non-drip paints contain polyurethane.

Emulsion paint has a lower VOC content than oil-based paint but can be more synthetic in composition, especially the low-VOC and low-odour types. It can be applied to woodwork, although it is not very durable used in this way. Much emulsion paint, however, is largely composed of acrylic, a plastic, which compromises the ability of underlying surfaces to breathe, while white emulsion may contain significant amounts of titanium dioxide, whose production causes a waste problem. However, emulsion paints can be removed from brushes using water rather than solvent-based brush-cleaning products.

Oil-based paint is a conventional finish for woodwork and metalwork because it dries to a hard durable finish that enhances water-resistance. However, oil-based paints, even those marketed as low-VOC, still have a higher VOC content than emulsion and should be avoided if possible. The solvents used in oil-based paints – white spirit or turpentine substitute – are synthetic ingredients derived from petroleum and are highly flammable.

OPPOSITE Star Yard in Norfolk, England, designed by Neil Winder, is both eco-friendly and built to accommodate climate change. The exterior is clad in grey-green larch, detailed to promote water runoff.

BELOW Solid timber external walls painted with lime colour, a natural and traditional finish.

LEFT Wooden walls painted with natural paints alleviate the 'woodiness' of all-timber construction. Other natural finishes for wood include penetrating oils such as linseed and tung oil, and beeswax.

RIGHT Natural paints are not as easy to apply as synthetic ones but are available in a good range of colours.

Natural paint contains solvents, binders and pigments derived from plant or mineral sources. The most common binder is linseed oil, which comes from the flax plant. Solvents include natural resin oils such as turpentine or citrus peel oils, which do, however, produce natural VOCs that can irritate or cause adverse reactions. In response, one manufacturer has recently brought out a range of water-based all-natural zero-VOC emulsion and gloss paints. Pigments include those derived from vegetable sources, such as madder root and oak bark, and crushed earth minerals. Vegetable pigments fade with time, but are ideal for soft colourwashed effects; mineral pigments are more intense and the colours last longer. Natural paints are not as easy to apply as synthetic ones and take longer to dry. They may require more coats for optimum coverage.

Milk or casein paint is a traditional paint whose main ingredients are lime, milk protein, and mineral pigment. The milk protein or casein is the binder, while the lime performs the same whitening function as titanium oxide. Milk-based paints produce a soft chalky finish; they can be used indoors or out and over any surface, although they are not advisable in damp conditions because they will flake off. Ready-made milk paint has a short shelf life; it is better to buy it in powdered form and mix with water to the required consistency, but take care as the lime in it is very caustic.

Distemper is another traditional type of chalk-based wall finish, typically applied over lime or clay plaster. It is not compatible with gypsum plasters and must be thoroughly washed off before renewing. Distemper is very soft and friable and tends to flake away from the surface.

Wood finishes include both those that are solvent- or plastic-based, for example polyurethane varnish, and various types of natural oils and waxes, which are preferable from the environmental point of view. Penetrating oils such as linseed oil and tung oil, applied in successive coats and buffed up, provide good water-resistance and have a pleasant odour. Beeswax is another natural finish for wood which scents the air and brings out the golden tones of the wood. To colour wood, apply a natural water-based wood stain.

Strategies for the environmentally friendly use of paints, varnishes and seals:

• Select the least harmful option: choose natural or casein paints over synthetic paints.

• Do without finishes wherever possible. Exposed brick does not need any finishing, while plasterwork can be left exposed and sealed with natural wax to prevent dusting.

• Avoid polyurethane seals and varnishes for interior wood surfaces and flooring. Use linseed oil, tung oil or beeswax instead.

• Keep rooms well ventilated during the application of paints or varnishes, wear protective clothing and masks and test samples before use to identify possible adverse reactions.

• Light-toned paint reflects light and so reduces dependence on artificial light sources.

• Lead is highly toxic but was once a common additive in paint. It is now banned in many parts of the world, including the United States, Britain and Australia. Old paintwork may contain lead so extreme care should be taken when stripping it away. You may need professional help.

IN PRACTICE

The old domestic virtues of thrift and household economy, practised by our grandparents, have an obvious role to play in living green. Making do and mending, home-growing produce for the table and keeping a check on wastage were second nature to generations for whom every penny counted. But eco practice is not about turning the clock back or about a self-denying deprivation: it entails rethinking our attitudes to possessions as much as reshaping habits and lifestyles.

The material false security of the last quarter-century, bought at the expense of environmental damage, has encouraged rampant consumer excess among the more affluent sectors in developed countries. Very little was disposable half a century ago; very little is not disposable nowadays. Major purchases such as cars, appliances and large items of furniture, are no longer acquired in the expectation that they will last; even without the technical and functional obsolesence built in by manufacturers, many products have a shelf life increasingly dictated by the whirlwind trends of the fashion cycle. We may not need to replace the car or buy a new sofa, but we are encouraged to want to, and to define ourselves by what (and how much) we own, with the result that our possessions now threaten to possess us. And when possessions get the upper hand, far from making our lives easier and more pleasant, they bring us more vexacious and niggly chores, not to mention putting pressure on space: they require servicing and looking after; they need to be kept somewhere; and, when we have finished with them, they need to be disposed of safely and without harm to the environment. The recent fad for minimalist design may well be seen as a symptom of our increasingly queasy relationship with our possessions.

Allied to this is the modern preoccupation with labour-saving and convenience: 'miracle' cleaners and detergents that take the elbow grease out of routine maintenance; gadgets and appliances of all descriptions, from leaf-blowers to ice-cream makers; one-stop shops and hypermarkets where all purchases can be made under one roof.

The environmental impact of the throwaway, convenience lifestyle has been well-documented: huge mountains of waste, excess consumption of energy and resources, harmful pollutants in our water and in the air, unsustainable reliance on the car, among others. But how convenient, labour-saving or ultimately satisfying is this way of living? Is it really more difficult to clean windows with vinegar and newspaper than with a proprietary chemical spray and a disposable cloth? Is it not faster to squeeze oranges by hand than pop them into a juicer and have to clean the appliance afterwards?

Even when tasks take more time or more effort performed the long way round, there may well be net gains in terms of personal satisfaction and health. Sedentary living, where even the shortest journeys are made by car and where every household chore must have its accompanying gadget, means that more people are turning to health clubs and exercise classes in an effort to maintain a basic level of fitness. And paradoxically, the short-term obsession with saving time has sped life up to such an extent that there barely seems time for anything any more.

By taking the long view, and sometimes the long way round, green living practices offer a way of reconnecting with basic rhythms and, through cooperation, with other people and the wider community.

CONSERVATION

Waste not, want not. Simple strategies for saving energy and conserving water can dramatically cut your consumption – and your bills – with a corresponding impact on natural resources. Few people set out to be wasteful; in most cases, leaving lights burning or taps dripping is merely down to lack of attention or laziness. It does not take much effort to break such bad habits, and the benefits can be substantial. The following strategies can be implemented by practically every household without major disruption or exhorbitant expense.

Saving energy

Although most of the appliances on the market today, from boilers to washing machines, are much more energy-efficient than equivalent models were ten or fifteen years ago, our consumption of energy is not decreasing at anything like the rate required to halt global warming and climate change. This is partly because we have more appliances and machines in our homes today than ever before and we use them more often. But it is vitally important to make the connection between domestic energy consumption and environmental damage: it isn't someone else's problem. In Britain, for example, domestic households account for a quarter of the total carbon-dioxide emissions released into the atmosphere every year.

Most energy consumed in the home is for heating or cooling, so this is where the greatest reductions can be made. Over-heated or over-cooled rooms are not the healthiest environments – indoor air quality can be improved if you warm yourself up with an extra layer of clothing or cool a space with natural ventilation rather than simply reaching for the thermostat.

• Get to know which appliances are most energy-hungry. Motor-driven appliances such as vacuum cleaners use relatively low amounts of power, compared to heating appliances, such as fan or radiant heaters, or appliances that combine motors with heating elements, such as washing machines. Greedy appliances include: cookers, grills, dishwashers, kettles, air conditioners, clothes driers, washing machines and irons.

• Vast savings on your energy bill and a dramatic reduction in energy consumption can be achieved simply by improving insulation. Insulate roof spaces, exterior walls and ground floors. Add insulation around hot-water tanks and pipes. Improvements to insulation have the potential to cut energy needs by two-thirds (see pages 28–31).

• Install double or triple glazing (see pages 31 and 146).

• Turn your thermostat down by a few degrees – put on a jumper rather than turn up the heat – and heat rooms for shorter periods. Turning the thermostat down by only one degree centigrade saves 8 per cent on fuel bills.

• For greater temperature control, install thermostats in rooms or on individual radiators. Use programmable controls that follow your patterns of use rather than heat spaces unnecessarily (see pages 32–33). If you can only afford to install one thermostat, put it in the room you use most frequently.

• Reflect heat back into the room by placing foil behind radiators on external walls.

• Make sure heat from radiators is not blocked by large items of furniture. Never hang curtains over radiators or you'll send the heat straight out of the window.

• Draughtproof doors and windows with weather-stripping.

• Upgrade boilers to new energy-efficient models (see page 33).

• Avoid fan and convection heaters if at all possible.

• Use low-energy lightbulbs. Compact fluorescent bulbs last ten times longer and use 75 per cent less electricity than ordinary bulbs; over their lifetime, they keep half a tonne of carbon dioxide from entering the atmosphere.

• Turn off lights when you leave a room and fit sensors or time clocks so that lights do not burn when not needed.

• Switch electronic equipment off completely (so that the red standby button is not lit). Even on standby mode, a device such as a television still consumes as much as 80 per cent of the power it requires when in use.

• Switch to a 'green' energy provider. International bodies, such as Friends of the Earth (see page 175), can advise on companies that have policies on renewable investment and energy efficiency.

Choosing and using appliances

Electrical appliances, particularly cookers, refrigerators and freezers, consume the greatest amount of energy after heating/cooling systems. Although it may sound superficially wasteful, it really does make environmental and economic sense to upgrade major appliances wherever possible to new energy-efficient models.

In the United States, Britain and Europe, energy-efficient appliances can be identified by labels that provide energy-consumption ratings. Look out for 'Energyguide' labels in the United States; 'European Energy' or 'Energy Saving Trust' in Europe and Britain. Old refrigerators are also likely to contain CFCs, which damage the ozone layer, and so they must be disposed of in an approved manner.

When it comes to smaller appliances and household gadgets, ask yourself if you really need to automate all routine tasks. Resist the temptation to buy a gadget such as a pasta maker or ice-cream maker you may only use a few times a year.

• Position fridges and freezers away from heat sources such as cookers, ovens, and dishwashers. A fridge next to a heat source takes up to 15 per cent more energy to run. They should also be positioned so that they are well-ventilated, with adequate air space at the rear and sides.

• Keep chest freezers in cool environments, such as basements, garages or unheated rooms so they don't have to work as hard.

• Don't over-chill a fridge or freezer; use eco-settings if available. Freezers work most efficiently when stocked to near capacity; fridges, however, should not be packed with perishables. Follow manufacturer's guidelines.

• Don't place warm food in a refrigerator; allow the food to cool to room temperature first.

• Similarly, don't leave the fridge door open for long periods when loading or unloading. Warm air causes a build-up of frost which affects the performance of the appliance.

• Larders and pantries located on the cool north-facing side of the house provide useful supplementary cool stores. If you have a larder, you may be able to make do with a small refrigerator.

• Use appliances such as washing machines and dishwashers at off-peak periods and run at full capacity.

• Line-dry clothing rather than use a tumble drier.

• Choose horizontal-axis drum washing machines, which use less energy, over tub varieties. Alternatively, use launderettes: shared machines are the most energy-efficient option.

• Choose low-temperature wash settings for all but the most heavily soiled clothes. Pre-soak heavily soiled clothes before washing.

• Gas cookers and stoves use half the energy of electric ones. If you must have an electric cooker, ceramic hobs are more efficient than rings.

• Convection ovens use less energy than conventional ones.

• Agas are very energy-efficient because they combine heat storage with a small heat source and provide background space heating.

• Microwave ovens cook food by radiation; long-term risks are still unknown and recommended safety levels vary widely from country to country. Older microwaves are more likely to leak radiation. Microwaves also increase reliance on packaged and convenience food, which leads to increased wastage. Use pressure cookers or other fast cooking methods instead.

• Don't overfill kettles and make sure you buy a model with an automatic cut-off switch.

• Service all appliances regularly to keep them running at maximum efficiency. Clean and defrost fridges, and keep filters and vents clear in washers and driers. Replace worn seals.

Saving water

In the developed world, we take more or less for granted the fact that every home has an endless supply of fresh drinking water. Yet a huge proportion of the world's population has no access to potable water at all and, with climate change, the situation is likely to worsen.

Water is a natural resource like any other, but we use it as if it were limitless, and without regard to the processes and treatments required to make it pure enough for domestic use. About half of domestic consumption is taken up by flushing toilets; a significant proportion simply trickles away through leaking valves, dripping taps or worn washers.

• Install a low-flush toilet. Early low-flush models were not very efficient, which actually led to an increase in water consumption, because people were flushing them twice. Now, however, designs have greatly improved. Whereas old-style toilets could use up to 20 litres per flush, new European recommendations set a maximum target of 6 litres. In the United States, a federal law phased in over a three-year period up to 1997 makes it compulsory for all new toilets to use no more than 6 litres per flush.

• If you don't want to install a low-flush toilet, you could simply flush less often, or displace a proportion of the water in the toilet cistern so that less is used per flush. One idea is to place a brick in the cistern; alternatively, use plastic bottles filled with pebbles or water.

• A variety of flow regulators are available which cut flows from taps and shower heads. Most are easy to fit.

• Take showers instead of baths (although power showers use a great deal of water).

• Choose dishwashers and washing machines that use less water and always run full loads.

• Avoid leaving taps running wherever possible: a running tap sends 9 litres of clean water down the drain per minute. Don't rinse dishes under a running tap or leave the tap on when shaving or brushing teeth. Wash cars using buckets of water rather than hose pipes.

• Garden irrigation accounts for a significant amount of water use, particularly in dry or affluent areas: in dry areas of the United States, for example, 80 per cent of a household's water consumption – that is, clean drinking water – goes on watering lawns, hosing down patios and paths, and cleaning cars. Collect rainwater in butts for garden watering and adopt water-saving gardening techniques (see p 42–43). You might also consider recycling greywater for garden irrigation.

• Water plants in the evening, rather than in the daytime.

• Fix all dripping taps; replace worn washers and cistern valves.

• Have a water meter fitted so you can monitor your consumption. Metering your supply can provide a pyschological incentive to cutting down on water use.

REDUCING WASTE

As fully paid-up members of the throwaway society, most of us barely stop to think about what happens to the rubbish we discard daily. The unpalatable truth is that a staggering proportion of domestic waste ends up either incinerated or dumped on landfill sites, disfiguring and rendering areas unproductive for generations to come. Although some countries have better track records than others in this respect – the Swiss, for example, recycle half their waste, compared to only 9.4 per cent in the case of Britain – excess waste remains a serious problem that needs to be addressed urgently.

There are three central strategies for reducing waste. The first is to reduce the amount we consume directly; the second is to re-use products or materials as often as possible; and the third is to recycle.

Use less

The T-shirt slogan 'I shop therefore I am' summarizes the unthinking albeit seductive patterns of consumption that have dominated contemporary society for the last several decades. For those serious about reducing waste, however, an important strategy is to attack the problem at its point of origin and buy only what you need (or indeed what you really want). This does not necessarily entail monastic self-denial. Buying fewer things means you can afford to buy better things – products of higher quality that will last longer and perform better. In many cases, such products repay care and maintenance and will go on looking good and functioning well for years.

Similarly, it is important to avoid products or materials that cannot be recycled or re-used. Avoiding plastics and plastic packaging can go a long way towards reducing waste.

Packaging accounts for a high proportion of domestic waste, particularly paper and plastic packaging in the form of clingfilm, polystyrene trays and plastic carrier bags. Of the 26 million tonnes of domestic waste generated in Britain every year, 3.2 million tonnes comprise packaging. Many goods are overpackaged; many have to be packaged to protect them during long periods of storage or transportation over great distances. To highlight the unnecessary wastage involved in food packaging, one group of British environmentalists staged a graphic demonstration in which shoppers removed all packaging from the goods they had just bought and left it behind at the supermarket checkout.

Some supermarkets have responded to such concerns by investigating ways of reducing packaging or by introducing recyclable or compostable packaging for organic produce. One retailer in Britain is experimenting with liquid-detergent vending machines, so customers can use refillable containers. Many now supply sturdy recyclable plastic carriers for a minimal sum; the bags can be re-used at least ten times and are replaced for free. In the United States, brown paper bags are more common than plastic carriers for groceries. And in some European countries, Holland, for example, supermarkets hand out only the flimsiest plastic bags so that shoppers are forced to bring their own carriers.

Strategies:

- Buy less; buy better-quality products that last longer. Good-quality products, particularly those in natural materials, often improve with use.
- Avoid products containing materials such as plastics which cannot easily be recycled and are not biodegradable.
- Buy goods that are as unpackaged as possible.
- Use re-usable shopping bags or baskets instead of plastic carrier bags. Refuse to have meat,

ice cream and other perishables double-wrapped: use an insulating bag instead.
• Buy staple goods in bulk to reduce packaging.
• Choose glass over plastic bottles.
• Reduce junk mail. Remove your name from direct mailing lists.

Re-use and repair

We often throw out things that could easily be re-used or repaired. Mending has become something of a lost art these days, but many items of furnishing or common household goods have decent lifespans if properly maintained. In this context, products made of natural or minimally processed materials are more long-lasting and easier to repair. Many synthetic materials, such as synthetic fibres and various types of plastic, do not acquire depth of character with age and use, but simply degrade unacceptably, which encourages us to throw them out sooner.

• Re-use glass jars and other containers as storage receptacles. Buy foodstuffs and staple products loose and decant. Store refrigerated food or leftovers in resealable containers rather than cover with cling film.

• Avoid disposable products such as plastic razors, plastic unrefillable pens, paper napkins, tissues, paper plates, cups and plastic cutlery.

• Eight million disposable nappies are thrown away in Britain alone every year. Use cloth nappies and a nappy-washing service, or at least alternate between disposables and cloth nappies. Instead of standard disposables, use non-gel, non-plastic ones which are biodegradable.

• Use rechargeable batteries, but make sure they contain no mercury or cadmium. Alternatively, use solar-powered equipment, such as solar-powered calculators or wind-up radios.

• Re-use envelopes.

• Repair and mend furniture and furnishings.

• Some cosmetic companies will refill bottles and other containers once the product has been used up.

• Unravel old woollen garments and re-use the yarn.

• Re-use worn-out bed linen or towels as cleaning rags.

• Re-use building materials, such as bricks, paving stones, timber and tiles, and individual building elements, such as windows, fireplaces, doors, fittings and fixtures.

• Children often grow out of clothing before it is worn out. Hand clothes down to friends and family or donate to charity shops.

• Upgrade electronic equipment wherever possible, for example, by installing a CD drive or additional memory into a computer. Dispose of old computers by donating to local schools or charities.

Recycling

Recycling schemes vary in effectiveness and availability, according to locality. Some types of recycling are more successful than others. Post-consumer plastic recycling, for example, still has a long way to go, chiefly because of the difficulty in sorting plastics into type. In some countries, separation of domestic waste to facilitate recycling is mandatory. New European legislation coming into effect in the next couple of years means that countries like Britain, which have a poor recycling record, will have to make drastic cuts in the amount of waste that ends up on landfill sites and improve local recycling schemes over the next two decades. Although bottle banks have become a common sight in many town centres, local authorities have been slow to implement doorstep collection schemes, which provide the strongest incentive for households to sort their domestic waste.

Recycling, however, particularly of common materials such as glass and metal, which form a large proportion of packaging, really does save energy. Making recycled aluminium uses 95 per cent less energy than aluminium made from scratch. Glass can be recycled with no loss of quality; recycled glass uses 32 per cent less energy to produce. Each tonne of recycled paper saves 2.2 cubic metres of landfill space and 15 trees.

Closed-loop recycling, where newspaper is recycled as newspaper or glass as glass, is considered to be preferable to downcycling, where a material is recycled into a material or product of inferior quality. In recent years, more innovative recycling schemes have taken a sideways step from the basic notion of consumption. In the United States, a number of companies now lend or lease their products – which include carpets and washing machines – to consumers, charging a fee for the period of use or per wash, for example, and undertaking to take the product back and recycle it at the end of its lifespan.

• Separate bins or recycling containers make it simpler to sort domestic waste by type: glass, metal, organic, paper. Glass should be sorted into clear, green and brown; remove corks and metal caps. Separate steel from aluminium cans by using a magnet (steel will stick to a magnet, aluminium will not). Kitchen foil, foil containers and foil bottle tops can also be recycled.

• Snip through plastic ring holders (the type used to hold six-packs) before you throw them away – they are a hazard to bird and marine life. Better still, buy cans loose.

• Make compost from organic kitchen waste. Compost can be made from peelings, any leftover food provided it is uncooked, as well as eggshells and bones.

• Waste-disposal units are eco-unfriendly because they direct waste into the water supply.

• Buy remanufactured printer cartridges and return your old cartridge in the envelope provided.

• Buy recycled paper products, such as stationery, toilet rolls and kitchen paper. Avoid pure white paper products; these have been bleached and bleach is a source of harmful dioxins.

• Buy salvaged building materials or elements; secondhand or antique furniture and furnishings.

• Think laterally for creative salvage approaches. At the simplest, this can be the old brick-and-scaffolding board approach to DIY shelving; look beyond for ways of incorporating panels, pallets, fittings and fixtures into fitted storage or furniture.

• Donate working fridges, washing machines, furniture or other bulky items to schemes that specialize in recycling to low-income families. Many charity shops take a range of goods, including books, ornaments, toys and clothes, and even unwanted reading glasses.

• Donate secondhand clothing and shoes to charities such as UK-based Traid, which uses the funds raised from their sale to combat world poverty.

Disposing of hazardous or non-recyclable waste

The best strategy overall in the case of hazardous waste is to avoid buying products that contain harmful chemicals or components in the first place or, if this is not entirely possible, to reduce the amount you consume. The disposal arrangements for hazardous waste vary widely from area to area: contact your local council or authority for details of schemes.

• Some local organizations will extract harmful CFC gases from old fridges.

• Some garages will accept defunct lead acid batteries.

• Never pour old motor oil down a drain or throw it out with the rubbish. Take it to a recycling centre.

• Some types of mobile phone can be recycled or reclaimed for parts. One British supermarket will recycle mobile phones, donating £5 per phone to charity.

• Low-energy fluorescent tubes can pose a disposal problem as they contain toxic heavy

metals. One supplier guarantees to take back old tubes and recycle them at the end of their life.
• Paints, solvents and garden chemicals should never be thrown away with domestic waste. Contact local authorities for details of hazardous-waste collection centres.
• Use up chemically based products fully and keep tightly sealed until finished.
• Use natural or alternative products whenever possible.

CLEANING AND MAINTENANCE

Many common household products, including cleansers, washing powders, furniture and floor polishes, air fresheners and insect-repellants contain a cocktail of chemicals harmful both to human health and the environment. Aerosol sprays used to contain CFCs but most are now air-propelled. However, few proprietary cleaners provide a full listing of ingredients, which means it is difficult to tell what chemicals you may be exposing yourself to. In many instances, it is easy to make a simple substitution of an eco product for a chemical or non-biodegradable one with no appreciable loss of performance. Just as effective, and often cheaper, are home remedies using common and harmless ingredients such as vinegar and baking soda.

Household products are heavily marketed in terms of 'power' and ease of use, promising instant cleaning and maintenance with next to no effort. However, unless dirt, grime or grease is allowed to build up over a period of time, routine cleaning really does not require superhuman effort; for example, it is far better to wipe oven and hob surfaces clean on a regular basis than be forced to blast off baked-on spills and splashes with an 'instant' toxic cleaner once in a blue moon.

There is also an increased tendency nowadays for people to over-clean, deodorize and bleach their homes, pursuing every last bacterium with missionary zeal, and manufacturers of antibacterial cleaning products have done their part to fuel such anxiety. Exposure to common bacteria – our grandmothers' 'peck of dirt' – has been shown to be an important means of boosting our natural immune systems. This is not to advocate domestic squalor, merely to point out that our homes do not need to be maintained at the level of hygiene one would expect in a surgical ward.

Eco cleaners and alternative products

• Choose eco cleaners over chemically-based products. Eco products are labelled with a complete list of ingredients. Cleaning agents may be based on coconut or palm oil; citrus oils provide scent; chamomile is a common softener. Eco cleaners are biodegradable and do not contain phosphates – a major component of ordinary washing powders and liquids, and a major pollutant.
• Read labels carefully when selecting shampoos, deodorants, soaps and other personal-care products. 'Natural' may not mean eco. Look out for products made from organic plant extracts.
• Heat and vaporize essential oils, such as lavender, sandalwood and ylang ylang to scent the air, rather than use a chemically based air freshener.
• Eco cloths with special dust- and dirt-trapping weaves, and which can be washed and re-used time and time again, can obviate the need for detergents and polishes.
• 'Eco balls' ionize washing water, which means rinsing is not required, thus saving water and electricity.
• Beeswax and linseed oil make good polishes and seals for wooden surfaces and furniture.
• Use hot water or a scraper to de-ice windscreens, rather than a de-icing spray.

Home remedies

Many common store-cupboard ingredients, such as vinegar and baking soda, make excellent cleansers and stain removers. Revisit old household manuals for ideas.

Vinegar

• Pour vinegar down the toilet and leave overnight to clean. Scrub well.

• Add a teaspoon of white vinegar to the wash to remove odours.

• Use vinegar and newspaper to clean windows.

• Warm white vinegar and salt can be used to clean copper and brass.

• Rinse hair in a little neat vinegar to prevent dandruff and remove soapy residue.

• Mix vinegar half and half with water to cleanse skin of soap.

Baking soda

• Add a teaspoon of baking soda to the wash to remove odours.

• Scour sinks and tiles with a paste made of baking soda and water. Baking soda can also be used as a toilet cleaner.

• Baking soda can be used dry as a deodorant.

• Place an opened box of baking soda in the fridge to absorb smells.

• Baking soda and salt can be used as an oven cleaner.

• To clean silver, line a tray or pan with aluminium foil, place the silver in, and add boiling water with baking soda and salt.

• Clean drains by flushing through with boiling water mixed with a quarter cup of baking soda and 50ml vinegar.

Borax

• Pre-soak heavy soiled clothing in a solution of borax and water (1tbsp to 4.5 litres).

• Use borax neat on a damp cloth to scour bathroom surfaces; alternatively scour surfaces with half a lemon dipped in borax.

• Borax is a natural disinfectant. Use half a cup of borax to 4.5 litres of water.

Insect-repellants

Insecticides found in fly- and ant-sprays, flea treatments for pets and mothballs, are highly toxic, and should be used only as a last resort. Many herbs, however, are natural insect-repellants.

• To repel moths, first of all keep clothes well aerated. Don't hang too closely on rails or in closets and don't overfill drawers. Use natural moth-repellants such as cedar chips, lavender and sachets of moth-repelling herbs; tuck among clothes or slip in pockets.

• Flies are deterred by the strong scent of cloves, or of herbs such as rue, rosemary, thyme and basil.

• Hot spices such as chilli or paprika sprinkled on the floor near doors act as an ant-repellant. Alternatively, if you can find the nest, try pouring boiling water into it, or sprinkle borax around common routes.

• Headlice, every parent's nightmare, are an increasing problem in many areas and are becoming resistant to the chemicals found in common proprietary treatments. The most effective way of controlling headlice is to check your child's hair regularly and thoroughly using a nit comb. Alternatively, or additionally, use tea-tree oil shampoo; tea-tree oil is a natural repellant.

• Bathe pets using tea-tree oil shampoo to deter fleas.

Garden pests

Domestic gardeners use more pesticides per hectare than farmers, which has serious implications not only for the environment in general, but also for human health. There are a wide range of strategies that can be employed to deter garden pests, which range from adopting old remedies dating from the times before chemical controls were available, to the type of plants you choose and the way you plant them. Organic gardening using biological controls also means adopting a new attitude to outdoor spaces, one which tolerates less-than-perfect specimens in the interests of better human and global health.

• Choose plants that are naturally resistant to pests and diseases.

• Feed the soil with organic compost to deter pests and diseases.

• Learn about natural predators of common garden pests. Insects such as spiders, ladybirds, hoverflies and dragonflies are good natural predators, so are birds, toads and hedgehogs.

• Strong-smelling plants, such as marigolds, onions and garlic, are natural insect-repellants.

• Mixed planting, rather than beds devoted to swathes of the same plant, helps to minimize decimation by predators.

• Shallow dishes sunk into the earth around the base of plants and filled with beer serve as slug traps.

• Powdery or granular matter, such as sawdust, soot and ash, placed around the base of plants can serve as a barrier to pests.

• A weak solution of washing-up liquid can be sprayed onto plants to deter greenfly. And it won't harm the ladybirds which are natural predators.

HOME AND AWAY

Eco living does not begin and end at the front door. Our dependence on the car and the long-distance transportation of food and other common household necessities incurs enormous costs, not merely in terms of energy consumption and the depletion of natural resources, but also on the character of our cities and communities.

In this context, acting local means rethinking our lifestyles to reduce journeys made by car as well the number of products acquired from far-flung places. Centralized supermarket distribution and the importing of foodstuffs means that the ingredients for a typical meal may have travelled tens of thousands of kilometres before they reach your plate. Unfortunately, this is particularly true of organic produce, which is often imported because local suppliers are as yet unable to meet demand.

If food travels far, so do most of us these days, journeying thousands of kilometres routinely every year on business and family holidays. Tourism is the biggest industry in the world, yet it is often left out of people's environmental considerations. It impacts on the environment in many ways, both direct and indirect. Directly, there is the impact of the vast amounts of aviation fuel consumed jetting passengers from one side of the globe to the other. Indirectly, there is the impact on indigenous communities, cultures and landscapes, not only from resort development but also through transport infrastructure, for example the building of highways and airports.

Getting about

• Walk, cycle or take public transport as much as possible. It is estimated that if as few as 1 per cent of American motorists left their car at home for a day each week, 132 million litres of petrol would be saved a year, keeping over 400 million kilograms of carbon dioxide from entering the atmosphere. Campaign for safe pedestrian and cycle routes.

• Form a car pool for essential commuting to school or office. If you don't use your car very often, consider giving it up and taking taxis instead. When you take into account road tax, servicing, insurance, depreciation as well as fuel costs, taxis or hiring a car for the occasional weekend may well be cheaper.

• Make sure you keep your car serviced and tuned so that it runs at maximum efficiency. Keep tyres pumped up: underinflated tyres waste 5 per cent of a car's fuel. Radial tyres improve petrol consumption.

• If you must buy a new car, research the market thoroughly to ensure you purchase a model with the best environmental credentials in terms of energy efficiency and emissions. Automatic cars are heavier on fuel than those with manual transmission. Another consideration is the recyclable content of a particular model.

• Use lead-free petrol.

• Cut your speed and avoid sudden acceleration and braking. Driving at 80kph uses 30 per cent less fuel than at 100kph.

On holiday

• Choose eco-friendly destinations for holidays. Eco tourism supports indigenous communities without harming the environment.

• Be conscious of precious resources, in particular water. Many holiday destinations are in areas where water is in short supply.

• Avoid buying souvenirs made of materials such as teak or ivory, to ease pressure on endangered species of flora and fauna.

• Don't routinely go abroad. Holidays taken at home often mean you can use trains rather than planes to reach your destination.

Food shopping

• Shop at farmers' markets or other local outlets where you can identify the provenance of foodstuffs and other goods.

• Buy organic, fairtrade products wherever possible. Buy fresh food that is in season, rather than fruit and vegetables airfreighted thousands of miles.

• Grow your own food in the garden or allotment.

• If you wish to shop at supermarkets, order by phone or on-line and have your groceries delivered. A delivery van can hold many separate orders, saving two car journeys for each one.

Green savings

• Invest ethically in 'green' pensions, insurance policies, mortgages and other financial products. Ethical investment is the fastest growing area of the stock market. Ethical investment or SRI (Socially Responsible Investment) funds invest in companies with good environmental and social policies.

CASE STUDY ARCHITECTS

Clare Design
Kerry & Lindsey Clare
41 McLaren Street
North Sydney
New South Wales 2060
Australia
Tel: +61 299 29 00 72
Fax: +61 299 59 57 65

Cole Thompson Associates
The Old Chapel
1 Holly Road, Twickenham
Surrey TW1 4EA
Tel: +44 20 8744 4450
Fax: +44 20 8744 4444
www.colethompson.co.uk

Cutler Anderson Architects
135 Parfitt Way SW
Bainbridge Island
Washington 98110
USA
Tel: +1 206 842 4710
www.cutler-anderson.com

Jersey Devil Architect/Builders
C/o Department of Architecture
University of Washington
Box 355720
Seattle
Washington 98195
USA
Tel: +1 206 543 7144
Fax: +1 206 543 2463

Jones Studio
4450 North 12th Street
Suite 104
Phoenix, Arizona 85014
USA
Tel: +1 602 264 2941
Fax: +1 602 264 3440
Email: maria@jonesstudioinc.com

Rick Joy Architect
400 South Rubio Avenue
Tucson, Arizona 85701
USA

TEL: +1 520 624 1442
Fax: +1 520 791 0699
www.rickjoy.com

Landström Arkitekter
Alsnögatan 12
116 41 Stockholm
Sweden
Tel: +46 8 679 90 60
Fax: + 46 8 611 82 52
www.landstrom.se

Glenn Murcutt & Associates Pty Ltd Architects
176a Raglan Street
Mosman
New South Wales 2088
Australia
Tel: +61 299 69 77 97

Gabriel & Elizabeth Poole Design Company
P O Box 1158
Noosaville DC
Queensland 4556
Australia
Tel: +61 754 42 45 33
Fax: +61 754 74 48 11
Email: poole@universal.net.au

David Sheppard Architects
49 Fore Street
Plympton St Maurice
Plymouth
Devon PL7 3LZ
Email:
david@davidsheppard-architects.com

Seth Stein Architect
15 Grand Union Centre
West Row
London W10 5AS
Tel: +44 20 8968 8581
Fax: +44 20 8968 8591
www.sethstein.com

Stutchbury & Pape Architects
4/364 Barrenjoey Road
Newport
New South Wales
Australia

Tel: +61 299 79 50 30
Fax: +61 299 79 53 67

Chris Thurlbourne
Alt.itude Architecture
www.alt-itude.com
Email: chris.thurlbourne@a-aarhus.dk

Sarah Wigglesworth Architects
10 Stock Orchard Street
London N7 9RW
Tel: + 44 20 7607 9200
Fax: + 44 20 7607 5800
www.swarch.co.uk

Michael Winter
The Boundary House
Upper Cumberland Walk
Tunbridge Wells
Kent TN2 5EH
Tel: +44 1892 539 709

OTHER ARCHITECTS WHOSE WORK IS FEATURED

Aantjes (Duurzaam Huis Leidsche Rijn)
Postweg 5
3941 KA Doorn
Tel: +31 343 416611

Obie Bowman
P O Box 1114
Healdsburg
California 95448
USA
Tel:/Fax: +1 707 433 783
www.sonic.net/~ogb

Chris Cowper
The Barn
College Farm
Whittlesford
Cambridgeshire CB2 4LX
Tel: +44 1223 835998
Fax: +44 1223 837327

Bill Dunster Architecture
Hope House
Molember Road
East Molesey, Surrey KT8 9NH

Tel: +44 20 8339 1242
Fax: +44 20 8339 0429
www.zedfactory.com

Forever Green
3 Onslow House
Castle Road, Tunbridge Wells
Kent TN4 8BY
Tel:/Fax: +44 1892 614300

Future Systems
The Warehouse
20 Victoria Gardens
London W11 3PE
Tel: +44 20 7243 7670
Fax: +44 20 7243 7690
www.future-systems.com

Mika Karkulahti
Tel: +358 9 451 5273
Fax: + 358 9 451 3015
Email: mike.karhulahti@hut.fi

Solar Century
91-94 Lower Marsh
London SE1 7AB
Tel: +44 20 7803 0100
Fax: +44 20 7803 0101
www.solarcentury.co.uk

Robert & Brenda Vale
Tel: +44 1636 815412

Neil Winder Architect
Star Yard
Millway Lane
Palgrave, Diss
Norfolk IP22 1AD
Tel: +44 1379 641592

Andrew Yeats
Eco Arc Architects
Old Village School
Harton
York YO60 7NP
Tel: +44 1904 468752
Fax: +44 1904 468492
Email: ecoarc@cwcom.net
www.ecoarc.co.uk

STOCKISTS AND SUPPLIERS

Organizations

Association for Environment-Conscious Building (AECB)
Nant-y-Garreg Farm
Saron
Llandysul
Carmarthenshire SA44 5EJ
Tel:/Fax: +44 1559 370908
www.aecb.net
Publishes magazine

Centre for Alternative Technology
Machynlleth
Powys SY20 9AZ
Tel:/Fax: +44 1654 703409
www.cat.org.uk
Publications and demonstration centre

Duurzaam Huis Leidsche Rijn
Johanniterpad 1
3544 VA Utrecht
Tel: +31 30 2412496
Email: info@duurzaam-huis.nl
www.duurzaam-huis.nl
Demonstration house open to the public, Friday 10.00–17.00 and Saturday 11.00–16.00

Environmental Building News
28 Birge Street
Brattleboro
Vermont 05301
Tel: +1 802 257 7300
www.BuildingGreen.com
US newsletter

The Findhorn Foundation
The Park
Findhorn
Forres
Morayshire IV36 0TZ
Tel: +44 1309 690154
Fax: +44 1309 691387
Tours of eco village available

Friends of the Earth
26-28 Underwood Street
London N1 7UJ
Tel: +44 20 7490 1555

Greenpeace
www.greenpeace.org
Information on pollution, solar energy, pvc and other environmental campaigns

International Association for Ecological Design
PO Box 27
S-2300 Svedala
Sweden
Tel: +46 40 40 48 32

The Soil Association
86 Colston Street
Bristol BS1 5BB
Tel: +44 117 929 0661
Fax: +44 117 925 2504

Walter Segal Self Build Trust
www.segalselfbuild.co.uk

Energy

American Aldes Ventilation Corp
4537 Northgate Court
Sarasota
Florida 34234 2124
Tel: +1 941 351 3441
Fax: +1 941 351 3442
freephone 800 255 7749
www.americanaldes.com
Manufacturers of heat-recovery ventilators

American Solar Energy Society
2400 Central Avenue, G-1
Boulder
Colorado 80301-2843
Tel: +1 303 443 3130

AstroPower
Solar Park
461 Wyoming Road
Newark
Delaware 197167-2000
Tel: +1 302 366 0400
Fax: +1 302 368 6474
freephone 800 800 8727
www.astropower.com
Large manufacturer of photovoltaic modules

Atlantis USA
4610 Northgate Bvld, Ste 150
Sacramento
California 95834
Tel: +1 916 920 9500
Fax: +1 916 927 1697
www.atlantisenergy.com
Manufacturer of solar slates

Bergey Windpower Co., Inc
2001 Priestley Avenue
Norman
Oklahoma OK 73069
Tel: +1 405 364 4212
Fax: +1 405 364 2078
freephone 888 669 6178
Manufacturer of small-scale wind turbines

BP Solar
989 Corporate Bvld
Llinthicum
Maryland 21703
Tel: +1 410 981 0240
Fax: +1 410 981 0278
www.bpsolar.com
Largest manufacturer of photovoltaic modules in the world

British Wind Energy Association
26 Spring Street
London W2 1JA
Tel: +44 20 7402 7102
www.bea.com

Center for Renewable Energy Technology
1200 18th Street NW no 900
Washington DC 20036
Tel: +1 202 530 2202
Fax: +1 202 887 0487

Ecohometec UK Ltd
22/24 Scot Lane
Doncaster DN1 1ES
Tel: +44 1302 769769
Fax: +44 1302 323323
Energy-efficient condensing boilers

Energy Saving Trust
21 Dartmouth Street
London SW1H 9BP
Tel: +44 20 7222 0101
Fax: +44 20 7654 2444
www.lightswitch.co.uk

NEF Renewables
The National Energy Foundation
Davy Avenue
Knowhill, Milton Keynes
Buckinghamshire MK5 8NG
Tel: +44 1908 665555
Fax: +44 1908 665577
www.natenergy.org.uk
Advice and contact details for suppliers and installers of renewable energy

Passive Solar Industries Council
1511 K St NW
Washington DC 20005
Tel: +1 202 628 7400

Solar Century
Unit 5, Sandycombe Road
Richmond, Surrey TW9 2EP
Tel: +44 870 735 8100
Fax: +44 870 735 8101
Suppliers of solar technology, including photovoltaic panels and solar slates

Solar Energy Industries Association
1616 H St NW, 8th floor
Washington DC 20006
Tel: +1 202 628 7745
Fax: +1 202 628 7779
www.seia.org
National trade association of companies providing solar products

Solar Sense
The Environment Centre
Pier Street
Swansea SA1 1RY
Tel: +44 1792 371690
Fax: +44 1792 371390
*Suppliers of solar products,
photovoltaics, wind-powered
systems*

SunEarth Inc
4315 Santa Ana St
Ontario
California 91671
Tel: +1 909 605 5610
Fax: +1 909 605 5613
freephone 800 776 5270
www.sunearthinc.com
Solar water-heating equipment

Tarm USA, Inc
5 Main St
PO Box 285
Lyme
New Hampshire 03768
Tel: +1 603 795 2214
Fax: +1 603 795 4740
www.woodboilers.com
*Manufacturer of multi-fuel
boilers and domestic hot-water
heaters*

Insulating materials

**Cellulose Insulation
Manufacturers Association**
136 S Keowee St
Dayton
Ohio 45402
Tel: +1 937 222 2462
Fax: +1 937 222 5794
freephone 888 881 2462
www.cellulose.org
*Association of American
cellulose insulation producers*

Excel Industries Ltd
13 Rassau Industrial Estate
Ebbw Vale
Gwent NP3 5SD
Tel: +44 1495 350655

Fax: +44 1495 350146
*Information about 'breathing
wall' construction; cellulose
insulation*

Klober Ltd
Pear Tree Industrial Estate
Upper Langford
N Somerset BS40 7DJ
Tel: +44 1934 853224
Fax: +441934 853221
Wool insulation

Water and waste

*Contact your local water
board/utility for information on
water meters and water-saving
devices*

American Standard
1 Centennial Way
Piscataway
New Jersey 08855
Tel: +1 800 223 0068
freephone 800 524 9797
www.us.amstd.com
*Manufacturers of gravity-flush
toilets*

Athena
17175 S.W. TV Highway
Aloha, Oregon 97006
Tel: +1 503 356 1233
Fax: +1 503 356 1253
freephone 888 426 7383
www.athenacfc.com
*Suppliers of controllable flushing
device to convert standard toilet
into low-flush toilet*

Bismart Distributors, Inc
8584 145 A St
Surrey
British Columbia, Canada V3S 2Z2
Tel: +1 604 596 5894
Fax: +1 888 663 4950
www.envirosink.com
*Suppliers of kitchen sinks that
drain to greywater system*

Cistermiser Ltd
Unit 1 Woodley Park Estate
59-69 Reading Road
Woodley
Reading Berkshire RG5 3AN

Tel: +44 1734 691611
Fax: +44 1734 441426
*Manufacturers of water-
conserving devices for cisterns*

Clivus Multrum, Inc
15 Union Street
Lawrence
Maine 08140
Tel: +1 978 725 5591
Fax: 978 557 9658
freephone 800 425 4887
www.clivusmultrum.com
*Manufacturer of composting
toilets*

Kingsley Clivus
5-7 Woodside Road
Eastleigh
Hants SO50 4ET
Tel: +44 1703 615680
Fax: +44 1703 642613
*Suppliers of low-flush and
composting toilets*

Niagara Conservation Corp
45 Horsehill Road
Cedar Knolls
New Jersey 07927
Tel: +1 973 829 0800
Fax: +1 973 829 1400
freephone 800 831 8383
*Suppliers of wide range of
water-conserving showerheads,
toilets and taps*

Green roofs

Erisco-Bauder Ltd
Broughton House
Broughton Road
Ipswich
Suffolk IP1 3QS
Tel: +44 1473 257671
Fax: +44 1473 230761

Roofscapes Inc
7114 McCallum Street
Philadelphia
Pennsylvania 19119
Tel:/Fax: +1 215 247 8784
www.roofmeadow.com

Gardening and landscape design

British Wild Flower Plants
31 Main Road
North Burlingham NR13 4TA
Tel:/Fax: +44 1603 716615

Ernst Conservation Seeds
9006 Mercer Pike
Meadville
Pennsylvania 16335
Tel: +1 814 336 2404
Fax: +1 814 336 5191
www.ernstseed.com

The Reveg Edge
PO Box 361
Redwood City
California 94064
Tel: +1 650 325 7333
Fax: +1 650 325 4056
www.ecoseeds.com

Materials: general

**Building for Health -Materials
Center**
PO Box 113
Carbondale
Colorado 81623
Tel: +1 970 963 0437
Fax: +1 970 963 0437
freephone 800 292 4838
www.
greenbuilder.com/CR/CedarRose.html
*Nationwide supplier of healthy,
eco products; catalogues
available*

Construction Resources
16 Great Guildford Street
London SE10HS
Tel:/Fax: +44 20 7450 2211

www.ecoconstruct.com
Building materials and products

Ecomerchant
The Old Filling Station
Head Hill Road
Goodnestone
nr Faversham
Kent ME13 9BY
Tel: +44 1795 530130
Fax: +44 1795 530430
www.ecoproducts.co.uk

Environmental Building Supplies
1331 NW Kearney St
Portland
Oregon 97209-2808
Tel: +1 503 222 3881
Fax: +1 503 222 3756
www.ecohaus.com
Flooring, FSC-certified wood,
tiles, furniture, finishes

Environmental Construction
Outfitters of NY
190 Willow Ave
Bronx
New York 10454
Tel: +1 718 292 0626
Fax: +1 718 401 4716
freephone 800 238 5008
www.environproducts.com

Natural Building Technology
Cholsey Grante
Ibstone
High Wycombe
Buckinghamshire HP14 3XT
Tel: +44 1491 638911
Fax: +44 1491 638630
www.natural-building.co.uk

Planetary Solutions
2030 17th St
Boulder
Colorado
Tel: +1 303 442 6228
Fax: +1 303 442 6474
freephone 800 488 2089
Cork, linoleum, wool, recycled
plastic carpet, reclaimed and

certified wood, bamboo,
recycled glass tile, natural
finishes

The Green Shop
Holbrook Garage
Bisley, Stroud
Gloucestershire GL6 7BX
Tel:/Fax: +44 1452 770629
Finishes, solar- and wind-
powered equipment; catalogue
available

Timber

Center Mills Antique Floors
PO Box 16
Aspers
Pennsylvania 17304
Tel: +1 717 334 0249
Fax: +1 717 334 6223
Salvaged and remilled wood
products, including flooring

Ecological Trading Co
659 Newark Road
Lincoln LN6 8SA
Tel: +44 1522 501850
Fax: +44 1522 501841
Supplier of certified timber

Forest Stewardship Council: UK
Unit D, Station Building
Llanidoes
Powys SY18 6EB
Tel: +44 1686 431916
Fax: +44 1686 412176
International organization that
sets standards for timber
management and products
worldwide; provides information
on suppliers of FSC-certified
timber

Forest Stewardship Council: US
1134 29th St NW
Washington DC 20007
Tel: +1 202 342 0413
Fax: +1 202 342 6589
freephone 877 372 5646
www.fscus.org

Harvest Forestry
1 New England Street
Brighton BN1 4GT
Tel: +44 1273 689725
Fax: +44 1273 622727
www.harvestforestry.co.uk

Straw bale

Straw Bale Building Association
Hollinroyd Farm, Butts Lane
Todmorden OL14 8RJ
Tel: +44 1706 818 216
www.
strawbalebuildingassociation.org.uk

Rammed earth

Adobe Factory
PO Box 519
Alcalde
New Mexico 87511
Tel: +1 505 852 4131

Centre for Earthen Architecture
School of Architecture
University of Plymouth
The Hoe Centre
Notte St
Plymouth PL1 2AR
Devon
Tel: +44 1752 233630
Fax: +44 1752 233634

CRATerre-EAG
BP 53
38092 Cedex
France
Tel: +33 474 95 64 21
Information service

Paper and straw

All Paper Recycling Inc
502 Fourth Ave, NW, Suite 7
New Prague
Minnesota 56071
Tel: +1 952 758 6577
Fax: +1 952 758 6751
Waste paper moulded into
building materials

Duro Sweden AB
Box 907
S–801 32 Gävle
Sweden
Tel: +46 26 65 65 00
Fax: +46 26 65 65 01
Environmentally friendly
wallapers

Isobord Enterprises Inc
1300 SW Fifth Ave, Suite 3030
Portland
Oregon 97201
www.isobordenterprises.com
Formaldehyde-free straw
particleboard

Bamboo

Bamboo Hardwoods Inc
3834 Fourth Ave
Seattle
Washington 98134
Tel: +1 206 264 2414
Fax: +1 206 264 9365
www.bamboohardwoods.com

Plyboo America Inc
745 Chestnut Ridge Road
Kirkville
New York 13082
Tel: +1 315 687 3240
Fax: +1 315 687 5177
www.plyboo-america.com

Plyboo (UK) Ltd
55-57 Main Street
Alford
Aberdeenshire AB33 8AA
Tel: +44 19755 63388

TimberGrass, LLC
9790 NE Murden Cove Dr
Bainbridge Island
Washington 98110
Tel: +1 206 842 9477
Fax: +1 206 842 9818
freephone 800 929 6333

Cork

Natural Cork, LLC
1710 North Leg Court
Augusta
Georgia 30909
Tel: +1 706 733 6120
Fax: +1 706 733 8120
www.naturalcork.com

Siesta Cork Tile Co
Unit 21, Tait Road
Gloucester Road,
Croydon, Surrey CR0 2DP
Tel: +44 20 8683 4055
Fax: +44 20 8683 4480

Wincanders Cork Flooring, Inc
586 Bogert Road
River Edge
New Jersey 07661
Tel:/Fax: +1 201 265 1407
freephone 800 828 2675

Linoleum

Armstrong DLW Commercial Floors
Centurion Court
Abingdon
Oxfordshire OX14 4RY
Tel: +44 1235 831296

DLW Linoleum
Armstrong World Industries
2500 Columbia Avenue
PO Box 3001
Lancaster
Pennsylvania 17604
Tel: +1 717 397 0611
freephone 877 276 7876
www.armstrong.com

Forbo Industries, Inc
Humboldt Industrial Park
Maplewood Drive
PO Box 667
Hazleton
Pennsylvania 18201
Tel: +1 570 459 0771

Fax: +1 570 450 0258
freephone 800 842 7839
www.forbo-industries.com

Forbo-Nairn Ltd
PO Box 1
Kirkaldy
Fife
Scotland KY1 2SB
Tel: +44 1592 643777

Rubber

US Rubber Recycling Inc
10440 Trademark St
Rancho Cucamonga
California

Dalsouple
PO Box 140
Bridgwater
Somerset TA5 1HT
Tel: +44 1984 667233

Stone

Bath and Portland Stone Ltd
Moor Park House
Moor Green
Corsham
Wiltshire SN13 9SE
Tel: +44 1225 810456

Delabole Slate
Pengelly
Delabole
Cornwall PL33 9AZ
Tel: +44 1840 212242

Paris Ceramics
583 Kings Road
London SW6 9DU
Tel: +44 20 7371 7778

Quartzitec
15 Turner Court
Sussex
New Brunswick
Canada E4E 2S1

Tel: +1 506 433 9600
Fax: +1 506 433 9610
freephone 877 255 9600
Stone tile made of quartz fragments bonded with Portland cement

Stonell Ltd
521/525 Battersea Park Road
London SW11 3BN
Tel: +44 20 7738 9990

Brick and tile

Bulmer Brick and Tile
The Brickfields
Bulmer
nr Sudbury
Suffolk CO10 7EF
Tel: +44 1787 269232
Fax: +44 1787 269040
Fired Earth
Twyford Mill
Oxford Road, Adderbury
Oxfordshire OX17 3HP
Tel: +44 1295 812088

Ibstock Building Products Ltd
21 Dorset Square
London NW1 6QE
Tel: +44 870 903 4013

The Mosaic Workshop
Unit B, 443-449 Holloway Road
London N7 6LJ
Tel: +44 20 7263 2997

Natural Tile
150 Church Road
Redfield
Bristol BS5 9HN
Tel: +44 117 941 3707
Fax: +44 117 941 3072

Terra Green Ceramics
1650 Progress Drive
Richmond
Indiana 47374

Tel: +1 765 935 4760
Fax: +1 765 935 3971
www.terragreenceramics.com
Tiles made from recycled aviation glass

Concrete and plaster

British Gypsum Ltd
East Leake
Loughborough
Leicester LE12 6JQ
freephone 0800 225225

Davis Colors
3700 E Olympic Boulevard
Los Angeles
California 90023
Tel: +1 323 269 7311
Fax: +1 323 269 1053
freephone 800 356 4848
www.daviscolors.com
Mineral pigments for self-colouring concrete flooring slabs

ECO-Block LLC
PO Box 14814
Fort Lauderdale
Florida 33302
Tel: +1 954 766 2900
Fax: +1 954 761 3133
freephone 800 595 0820
www.eco-block.com
Insulating concrete forms

Greenblock Worldwide Corp
PO Box 749
Woodland Park
Colorado 80866
Tel: +1 719 687 0645
Fax: +1 719 687 7820
freephone 800 216 1820
www.greenblock.com
Insulating concrete forms

Glass and windows

The Efficient Windows Collaborative
Alliance to Save Energy
1200 18th St NW Suite 900
Washington DC 20036
www.efficientwindows.org

Paramount Windows, Inc
105 Panet Rd
Winnipeg
Manitoba R2J 0S1
Canada
Tel: +1 204 233 4966
Fax: +1 204 231 1043
www.paramountwindows.com
Long-established producer of energy-efficient windows

Pilkington Glass Ltd
Prescot Road
St Helens
Merseyside WA10 3TT
Tel: +44 1744 629000
Fax: +44 1744 613049
Manufacturer of low-E glass

Swedish Window Co Ltd
Millbank
The Airfield
Earls Colne
Colchester
Essex CO6 2NS
Tel: +44 1787 223931
Fax: +44 1797 224400

Metal

SMI Steel Products
4365 Highway 278 W
PO Box 2099
Hope
Arkansas 71802
Tel: +1 870 722 6255
Fax: +1 870 777 1966
www.smisteelproducts.com
Manufacturer of 100% recycled resource-efficient steel beam

Steel Recycling Institute
680 Andersen Drive
Pittsburgh
Pennsylvania 15220 2700
www.recycle-steel.org
Association promoting recycling of steel products

Plastic

Smile Plastics Ltd
The Mansion House
Ford
Shrewsbury SY5 9LZ
Tel: +44 1743 850267
Fax: +44 1743 851067
Recycled plastic sheets for furniture, work surfaces and panelling

Yemm and Hart
1417 Madison, Suite 308
Marquand
Missouri 63655-9153
Tel: +1 573 783 5434
Fax: +1 573 783 7544
www.yemmhart.com
Recycled plastic sheets and panels

Natural fabrics and weaves

Crucial Trading
79 Westbourne Park Road
London W2 5QH
Tel: +44 20 7221 9000
Fax: +44 20 7727 3634
Natural fibre floor coverings, from coir and sisal to jute

Fired Earth (see entry under Brick and tile)

Natural fibre floor coverings

Greenfibres
Freepost LON 7805
49 Blackheath Road
London SE10 8BR
Tel: +44 20 8694 6918
Fax: +44 20 8694 1296

The Healthy House
Cold Harbour
Ruscombe
Stroud
Gloucestershire GL6 6DA
Tel: +44 1453 752216
Fax: +44 1453 753533
www.healthy-house.co.uk
Pure untreated cotton bedding and other products

Ian Mankin
109 Regents Park Road
London NW1 8UR
Tel: +44 20 7722 0997
Fax: +44 20 7722 2159
Natural fabrics

Paints, varnishes and seals

Auro Organic Paints Supplies Ltd
Unit 1
Goldstones Farm
Ashdon,
Saffron Walden
Essex CB10 2LZ
Tel: +44 1799 584888
Fax: +44 1799 584041

EarthTech
PO Box 1325
Arvada
Colorado 80001-9998
Tel: +1 303 465 1537
Fax: +1 303 465 5153
www.earthtechinc.com

Nutshell Natural Paints
PO Box 72
South Brent
TQ10 9YR
Tel: +44 1364 73801
Fax: +44 1364 73068

Old Fashioned Milk Paint Co
436 Main Street
PO Box 222
Groton
Maine 01450

Salvage

American Salvage
9200 NW 27th Ave
Miami
Florida 33147
Tel: +1 305 836 4444
Fax: +1 305 691 0001
www.americansalvage.com

Architectural Salvage Register
Hutton & Rostron
Netley House
Gormshall GU5 9QA
Tel: +44 1483 203221
Fax: +44 1483 202911
Register of architectural materials suppliers

LASSco
St Michael's
Mark Street
London EC2A 4ER
Tel: +44 20 7749 9944
Fax: +44 20 7749 9941
Long-established and comprehensive salvage

Whole House Building Supply
731-D Loma Verde Ave
Palo Alto
California 94303-4161
Tel:/Fax: +1 650 856 0634
freephone 800 364 0634
www.driftwoodsalvage.com

Page numbers in italic refer to the illustrations

ACKNOWLEDGEMENTS

1 Hans Pattist; 2 Arcaid/Simon Kenny/architect Ken Latona; 6 Alt.itude Arkitekttegnestue/Chris Thurlbourne; 9 Hans Pattist; 10 Arcaid/Simon Kenny/architect Ken Latona; 13 BRE/John Russell/architects Cole Thompson Associates; 14 ESTO/Scott Frances; 17 Richard Davies/architects Future Systems; 18 View/Dennis Gilbert/architect Bill Dunster; 20 Christoph Kicherer/Stutchbury & Pape Architects; 21 Ray Main/Mainstream/architect Jon Broome; 22–23 Ray Main/Mainstream/architect Chris Cowper; 24 Arcaid/Richard Glover; 25 Jerry Harpur/Munich Rooves; 26 Reiner Blunck/architect Glen Murcutt; 27 Christoph Kicherer/Stutchbury & Pape Architects; 29 World of Interiors/Tomasso Mangiola/architects Gerold Schneider & Katia Polletin; 30 Michael Lamotte; 31 Arcblue.com/Peter Durrant/architects Cole Thompson Associates; 32 Deidra Walpole/designer Van Atta Associates; 32–33 Solar Century; 34 Arcaid/Jeremy Cockayne/architects Eco Arc; 35–36 Reiner Blunck/architect Gabriel Poole; 37 Ray Main/Mainstream; 38-39 Richard Davies/architect Future Systems; 39 Axiom/Dave Young/architect Bill Dunster; 40 Richard Davies/architect Neil Winder; 41 Arcaid/Martine Hamilton-Knight/architect Robert & Brenda Vale; 42–43 Derek St Romaine/designer Daniel Lloyd-Morgan; 43 Charles Mann/designer Nancy Wagoner; 46–49 Paul Smoothy/Sarah Wigglesworth Architects; 50–51 Jefferson Smith/Sarah Wigglesworth Architects; 53–57 Undine Pröhl/architects Rick Joy; 58–63 Stephen Oxenbury/architect Glen Murcutt; 64–65 Arcblue.com/Peter Durant/architects Cole Thompson Associates; 66 BRE/John Russell/architect Cole Thompson Associates; 67 Arcblue.com/Peter Durrant/architect Cole Thompson Associates; 68–73 Timothy Hursley/architect Jones Studio Inc.; 74–75 Arcaid/Alan Weintraub/architects Jersey Devil; 76–81 Reiner Blunck/architect Clare Design; 82–85 Alt.itude Arkitekttegnestue/Chris Thurlbourne; 86–91 Ray Main/Mainstream/architect David Sheppard; 92–93 Lars Hallén/Design Press/architect Anders Landström; 94–99 View/Peter Cook; 101–103 Geoff Lung/Stutchbury & Pape Architects; 104–107 Undine Pröhl/architect James Cutler; 108–113 Richard Davies/architect Seth Stein; 115–119 Reiner Blunck/architect Gabriel Poole; 120 Richard Davies/architect Neil Winder; 123 Christoph Kicherer/Stutchbury & Pape Architects; 124 Ray Main/Mainstream; 127 Reiner Blunck/architect Glen Murcutt; 128 Ray Main/Mainstream/architect Jon Broome; 129 Arcaid/Simon Kenny/architect Ken Latona; 130 Tom Rider/architect Obie Bowman; 131 left Lassco Flooring; 131 right Arcaid/Alberto Piovanno/architect Edward Rojas Vega; 132 Lars Hallén/Design Press/architect Mika Karkulahti; 133 Michael Moran/architects Moneo Brock Studio; 134 Fired Earth; 135 above Duro Enfärgad; 135 below Ray Main/Mainstream; 136 Timbergrass/Joe Tschida; 137 www.elizabethwhiting.com/Rodney Hyett; 138 Forbo Nairn/Marmoleum; 139 left Forbo Nairn/Marmoleum; 139 right Crucial Trading Company; 140–141 Ray Main/Mainstream; 142–143 Marcella Echavarria/architect Luis Restrepro/Claudia Uribe Touri; 144 Craig Fraser/architects J Jacobson & R Tremeet/interior designed by Cheryl Cowley; 145 Michael Freeman; 147 Richard Sexton/architect Obie Bowman; 148 Edifice/Philippa Lewis/architect Proctor Matthews; 149 Reiner Blunck/architect Glen Murcutt; 150 Bar + Knell (Beata Bär, Gerhard Bär, Hartmut Knell) in cooperation with the Deutsche Gesellschaft für Kunststoff-Recycling mbH (DKR); 151 Smile Plastics/Colin Williamson; 153 The Interior Archive/Edina van der Wyck/designer Melin Tregwynt; 154 left Crucial Trading Company; 154 right Creative Communications; 155 Greenfibres; 156 Richard Davies/architect Neil Winder; 157 Lars Hallén/Design Press/architect Anders Landström; 158 Arcaid/Alberto Piovanno/architect Edward Rojas Vegas; 159 Arcaid/Richard Bryant/architect John Pardey; 164 Smile Plastics/Colin Williamson.

AUTHOR'S ACKNOWLEDGEMENTS

I would like to thank Helen Lewis, Nadine Bazar and all the team at Quadrille, particularly Hilary Mandleberg whose meticulous editorial support was hugely appreciated. Special thanks must also go to Robin Hillier of Forever Green who read the manuscript and made many helpful comments and suggestions.